ENCYCLOPAEDIA OF EMINENT THINKERS

VOLUME-XXV: ANNIE BESANT

Volumes by the same author

The Political Thought of ***Dadabhai Naoroji***
The Political Thought of ***Tej Bahadur Sapru***
The Political Thought of ***C. Rajagopalachari***
The Political Thought of ***Subhas Chandra Bose***
The Political Thought of ***Bhagat Singh***
The Political Thought of ***Annie Besant***
The Political Thought of ***C.R. Das***
The Political Thought of ***Lala Lajpat Rai***
The Political Thought of ***Sir Pherozeshah Mehta***
The Political Thought of ***Mahadev Govind Ranade***

About the Author

Dr. (Miss) Ashu Pasricha is a faculty in Gandhian studies at Panjab University, Chandigarh. She has many books to her credit including *Gandhian Approach to Integrated Rural Development; Public Administration: Discipline and Dimensions; Peace Studies: Discipline and Dimensions; WTO, Self-reliance and Globalisation: and Gandhi's Concept of State and Power.* Also, she has published a number of research papers in various Indian and foreign journals of repute.

ENCYCLOPAEDIA OF EMINENT THINKERS

VOLUME-TWENTY FIVE

THE POLITICAL THOUGHT OF ANNIE BESANT

Dr. Ashu Pasricha
Department of Gandhian Studies
Panjab University
Chandigarh

CONCEPT PUBLISHING COMPANY, NEW DELHI-110 059

Cataloging in Publication Data—DK
Courtesy: D.K. Agencies (P) Ltd. docinfo@dkagencies.com

Encyclopaedia of eminent thinkers.
p. cm. — (Gandhian studies and peace research series ; 11)
Vols. 11- lack series statement.
In Indian context.
Includes bibliographical references.
Includes indexes.
Contents: v. 25. The political thought of Annie Besant / Ashu Pasricha.
ISBN 13: 9788180695858
ISBN 10: 8180695859

1. Statesmen—India. 2. Political science—India. 3. India—Politics and government. I. Sharma, Jai Narain, 1951- II. Pasricha, ashu. 1967- III. Series: Gandhian studies and peace research series ; 11.

DDC 954.035 22

ISBN-13: 978-81-7022-684-0 (Series)
ISBN-10: 81-7022-684-8 (Series)
ISBN-13: 978-81-8069-585-8 (Vol. XXV)
ISBN-10: 81-8069-585-9 (Vol. XXV)

First Published 2009

Published and Printed by
Ashok Kumar Mittal
Concept Publishing Company
A/15-16, Commercial Block, Mohan Garden
New Delhi-110059 (India)

Phones: 25351460, 25351794
Fax: 091-11-25357103
E-mail: publishing@conceptpub.com

DEDICATED
TO
MY SISTER
DR. SEEMA PASRICHA
AND
NEPHEW
MASTER HEMANT MALHOTRA

PREFACE

Dr. Annie Besant was one of the celebrities of the modern world. India owes a special debt to her. She rendered very valuable services to the cause of education, the study of religion, and also to the cause of India's freedom.

The emergence of Annie Besant on the political scene of India was a significant phenomenon, as at that point of time the country stood in need of a leader, who could give a practical lead to the nation. There was sudden and unexpected dearth of prominent leaders in India. Gopal Krishna Gokhale died on 19 February 1915, and Pherozeshah Mehta followed him in November 1915. D.E. Wacha was fairly old and infirm. Surendranath Banerjea was not quite in tune with the new thought. Bal Gangadhar Tilak, who was released in June 1911, was physically shattered and it was just not possible for him to undertake strenuous whirlwind tours of various regions of India. Lala Lajpat Rai was unhappy with the state of affairs and was, at that time, in the U.S.A. Gandhi had just arrived in India and it took him sometime before he could introduce himself well to the Indian masses after the Lucknow session of the Indian National Congress in 1916.

Dr. Besant's active participation in Indian politics synchronised with the commencement of the First World War. For a period of 15 years thereafter she dominated the political scene, contributing and subsequently working out, with thoroughness and precision, two great conceptions, which all progressive forces in the country immediately accepted as of vital importance. These conceptions were first, Home Rule for India as a member of the British Commonwealth of Nations; and second, after the acceptance of the goal of responsible government by Mr. Lloyd George's Cabinet at the end of the war, the principle of self-determination should come into force for India—a Constitution framed for India by Indians.

For a foreigner to be endowed with the spirit and actuated with her longing for a great many years was a wonderful thing; but in addition she did a great deal to add to the culture and philosophy of India, to make religion better than it was. That she lived to be a great teacher in this country, and that hundreds of thousands heard her, read her writings, and were glad to learn of their own philosophy from her, is tribute to her greatness and her power, second to none that I can think of.

With this perspective in mind, the present work is an attempt to discuss the political ideas of Dr. Annie Besant.

Personal acknowledgments are no easy matters where so many have been kind. My special thanks are due to Professor R.N. Pal, Professor N. Radhakrishnan, Dr. K.S. Bharathi, Dr. A.D. Mishra, Professor Jai Narain Sharma, Dr. Meena Dutta, Dr. Seema Pasricha, Dr. Rashmi Sharma, Mr. Ajay Pasricha (Advocate) and Master Hemant Malhotra. Beyond that I can make only an all-embracing salute to the galaxy of the scholars whose works I have consulted.

I am also grateful to Shri Ashok Kumar Mittal, my publisher for his consistent zeal in the printing and publication of this volume.

Chandigarh **ASHU PASRICHA**

CONTENTS

Encyclopaedia of Eminent Thinkers Series

Vol. I. The Political Thought of Mahatma Gandhi/ *K.S. Bharti*
Vol. II. The Political Thought of Nehru/ *K.S. Bharti*
Vol. III. The Political Thought of Rabindra Nath Tagore/ *K.S. Bharti*
Vol. IV. The Political Thought of Radhakrishnan/ *K.S. Bharti*
Vol. V. The Political Thought of Vinoba/ *K.S. Bharti*
Vol. VI. The Political Thought of Aurobindo/ *K.S. Bharti*
Vol. VII. The Political Thought of Jayaprakash Narayan/ *K.S. Bharti*
Vol. VIII. The Political Thought of Vivekananda/ *K.S. Bharti*
Vol. IX. The Political Thought of Ambedkar/ *K.S. Bharti*
Vol. X. The Political Thought of M.N. Roy/ *K.S. Bharti*
Vol. XI. The Political Thought of Dadabhai Naoroji/ *Ashu Pasricha*
Vol. XII. The Political Thought of Veer Savarkar/ *Jai Narain Sharma*
Vol. XIII. The Political Thought of M.A. Jinnah/ *Jai Narain Sharma*
Vol. XIV. The Political Thought of Tej Bahadur Sapru/ *Ashu Pasricha*
Vol. XV. The Political Thought of C. Rajagopalachari/ *Ashu Pasricha*
Vol. XVI. The Political Thought of Subhas Chandra Bose/ *Ashu Pasricha*
Vol. XVII. The Political Thought of Mohammad Iqbal/ *Jai Narain Sharma*
Vol. XVIII. The Political Thought of K.B. Hedgewar/ *Jai Narain Sharma*
Vol. XIX. The Political Thought of Bhagat Singh/ *Ashu Pasricha*
Vol. XX. The Political Thought of M.S. Golwalkar/ *Jai Narain Sharma*
Vol. XXI. The Political Thought of Sir Syed Ahmed Khan/ *Jai Narain Sharma*
Vol. XXII. The Political Thought of Mahadev Govind Ranade/ *Ashu Pasricha*
Vol. XXIII. The Political Thought of Pandit Deendayal Upadhyaya/ *Jai Narain Sharma*
Vol. XXIV. The Political Thought of Lokmanya Bal Gangadhar Tilak/ *Jai Narain Sharma*
Vol. XXV. The Political Thought of Annie Besant/ *Ashu Pasricha*

Forthcoming

The Political Thought of C.R. Das/ *Ashu Pasricha*
The Political Thought of Gopal Krishna Gokhale/ *Jai Narain Sharma*
The Political Thought of Lala Lajpat Rai/ *Ashu Pasricha*
The Political Thought of Maulana Abul Kalam Azad/ *Jai Narain Sharma*
The Political Thought of Sir Pherozeshah Mehta/ *Ashu Pasricha*

1

LIFE AND TIMES

Annie Wood, as she was known till her marriage, was born on the lst October 1847 in London. Her mother was of pure Irish descent and her father was Irish on his mother's side though belonging to the Devonshire Woods on his father's. One of her ancestors had been Mayor of London; and her family gave England a Lord Chancellor in the person of Lord Hatherley. In essence, however, Annie Besant loved the Irish and Irish ways. She has remarked that "the Irish tongue is music in my ear and the Irish nature dear to my heart". Another remark of hers is characteristic, namely "Blessings on the quick tongues and warm hearts, on the people so easy to lead, so hard to drive". One of her aunts, she recounts, was very proud of her family tree and claimed descent from seven kings of France. This family descent gave her a training in reticence and the pride of honour which in her own words, was strange preparation for her stormy public life—a much attacked and slandered life. The very pride became a shield against degradation and she says in feeling language that "however lost my public reputation, I could never bear to become sullied in my own sight".[1]

Annie's father died when she was five years old. She says of him that, he was keenly intellectual and splendidly educated, a mathematician and a good classical scholar, a master of French, Italian, German, Spanish and Portuguese, with a smattering of Hebrew and Gaelic. It is worthy of remark, in view of her later development that her father, though he was a student of philosophy, was deeply and steadily sceptical. He had outgrown the orthodox beliefs of his days. Her mother was, however, deeply religious although, as she herself remarks, she put on one side as errors the doctrines of eternal punishment, vicarious atonement and the infallibility of the Bible.

Very remarkably, in her early days, Annie Besant took her religion most seriously. As her mother lay dying, she said to Annie, "My dear, you have always been too religious".[2] As a child, Annie was mystical and imaginative, seeing visions and dreaming dreams. Sensitiveness to impressions other than physical ones was, in fact, a marked feature in the family to which she belonged. Elves and fairies were very real to them.

The childhood of Annie Besant was a period of struggle and anxiety on the part of her mother on account of financial troubles. Mrs. Wood resolved that her son should be educated in Harrow, She carried out that scheme although she was practically penniless. She kept a boarding house and maintained herself and educated her son at Harrow on the proceeds. The task of looking after Annie was taken over by Miss Marryat, a sister of the famous novelist, Captain Marryat. Miss Marryat was a born teacher and she pioneered, in many ways, the training of children on the basis of the least pain and most enjoyment to themselves—a system analogous to that evolved by Madame Montessori, who later on co-operated with Annie Besant in the educational sphere.

Annie was brought up on the borders of Devon, Miss Marryat having taken a beautiful place, Fernhill, and started a Sunday school and Bible class there. This was a time of pervasive materialism in England. It was not the philosophic materialism of the few but the religious materialism of the many. Annie's education, however, was accented on the religious side and her favourite books were *The Pilgrim's Progress* and Milton's *Paradise Lost*. Curiously sensitive as she was, she says that she was pained because she could not look back to an hour of "conversion"; that she felt that her dreamy longings were very poor things compared with the vigorous "sense of sin" spoken of by the preachers.

When Annie was fourteen years old, Miss Marryat took her to Germany after she was duly instructed in the German language for some months. She was very strictly guarded; and, as she says, "dear auntie" looked on all young men as wolves to be kept far from her growing lamb. Later on, Miss Marryat and she spent some months in Paris during the hey-day of the Empire. Annie was greatly affected by the solemn beauty of Notre Dame, La Madelaine and St. Roch. Her religious nature was deeply aroused by the colour, fragrance and pomp in the religious services, and by the picture galleries of the Louvre crowded with Madonnas and Saints. As she says, she was not allowed to read love stories and poems and her days were filled with broodings over the days when girl martyrs were blessed with visions of the King of Martyrs, when sweet St. Agnes saw her celestial Bridegroom and Agnes stopped to whisper melodies in St. Cecilia's raptured ear.[3]

On her return to England, she perfected her French and German, came back to her mother and underwent a very thorough musical education. In the meantime, her mother was seriously afflicted in many ways; she had been systematically cheated by a lawyer whom she had trusted. But all the stern realities of life were kept back from her by her loving mother who guarded and shielded the childhood and youth of Annie from every touch of pain and anxiety. During the years preceding her engagement, she concentrated on a study of the writings of the Fathers of the early Christian Church. Looking back over these years, she makes a touching and accurate observation that the keynote of her life, through all the blunders and blind mistakes and clumsy follies, has been the longing for sacrifice as against self-interest. She later on regarded and referred to this as a tendency brought over from her previous life and dominating the present one.

In 1866, she was introduced to a clergyman, Rev. Frank Besant, a young Cambridge don, whom she married after a brief courtship. Remarkably, however, at the same time that she met Frank Besant, she began to encounter many doubts as to the Christian Gospels, which she recorded in her diaries. So far as her inner life was concerned, she was still filled with passionate religious fervour, and a philosophy which she characterises as really the human passion of love transferred to an ideal. Holding this view of religion, she idealised the clergy man as a special messenger and servant of the Lord. Her engagement to Frank Besant took place within a few weeks, and, as she says, she drifted into an engagement with a man she did not pretend to love. She tried to break the engagement but her mother dissuaded her and so she married when she was 20 years of age, with no more idea of marriage relations than if she had been four years old.

In 1867, she was introduced by Roberts, who was styled the "Poor Man's lawyer", to radical politics. Roberts was an admirer of the great orator and politician, John Bright, who threw himself into the Irish Fenian Movement. Roberts defended the Fenians but Justice Blackburn, the banging judge as he was called, dominated the trial of the cases. Mrs. Annie Besant states that her first experience of an angry crowd was on the day on which Roberts and she drove to the court. The Irishmen in the street muttered curses against the "d - d English who were going to see the boys murdered". The Jury that tried the case included a man who declared that he did not care what the evidence was but would hang every d - d Irishman of the lot. The verdict of guilty was given and the accused Larkin and O'Brrien, were hanged.[4]

It was in this year that Annie Besant first got to be acquainted with the work of Charles Bradlaugh and, as she says, she wrote with deep emotion the leading article dated 24th November in the *National Reformer* edited by him. It appeared under the caption "Where is our boasted English freedom?" In December 1867 commenced the crisis in Annie Besant's distressful married life. She observes : "We were an ill-mated pair, my husband and I, from the very outset; he, with very high ideas of a husband's authority and a wife's submission, precise, methodical, easily angered and with difficulty appeased; I, accustomed to freedom, indifferent to home details, impulsive, hot-tempered and proud as Lucifer."[5] As she herself states, another kind of treatment may have turned her into a fair imitation of a proper conventional article. As a result, she lived with a wall of ice around her within which she waged mental conflicts that nearly killed her. To relieve her solitude, she began writing short stories. These short stories were published and brought her a cheque which gave her intense delight, as it was the first money which she had earned, and the pride of earning it was added to the pride of authorship.

In 1869 a son was born and, later on in 1870, a daughter. The babies gave her a new interest and a new pleasure and she nursed them herself. In 1871, the two children got whooping cough and she states that she fought with death for her children. Once they were out of danger, Annie Besant collapsed physically and then rose, in her own language, to face a struggle which lasted for three years and two months—a struggle which transformed her from a Christian to an atheist. No one who has not felt it knows the anguish inflicted by doubt on an earnestly religious soul. The following passage from her Autobiography is significant and moving:

"It was the long months of suffering through which I had been passing, with the seemingly purposeless torturing of my little one as a climax, that struck the first stunning blow at my belief in God as a merciful Father of men. I had been visiting the poor a good deal, and had marked the patient suffering of their lives; my idolised mother had been defrauded by a lawyer she had trusted, and was plunged into debt by his non-payment of the sums that should have passed through his hands to others; my own bright life had been enshrouded by pain and rendered to me degraded by an intolerable sense of bondage; and here was my helpless, sinless babe tortured for weeks and left frail and suffering. The smooth brightness of my previous life made the disillusionment more startling, and the sudden plunge into conditions so new and so unfavourable dazed and stunned me. My religious past became the worst enemy of the present suffering. All my personal belief in Christ, all my intense faith in His constant direction of affairs, all my habit of continual prayer and of realisation of His Presence—all were against me now. The very height of my trust was the measure of the shock when the trust gave way."[6]

Conventional consolations and conventional prayers, at this time, seemed to be an aggravation of suffering. Her attitude led to grave quarrels between husband and wife. Her religious wretchedness increased the unhappiness of home life. But, in spite of all her outer troubles, she made up her mind to examine Christianity, to examine its dogmas, one by one, so that she could never again say "I believe" where she had no proof. The problems she dealt with were four:

(1) The eternity of punishment after death.

(2) The meaning of "goodness" and "love", as applied to a God who had made this world, with all its sin and misery.

(3) The nature of the atonement of Christ, and the "justice" of God in accepting a vicarious suffering from Christ, and vicarious righteousness from the sinner.

(4) The meaning of "inspiration" as applied to the Bible, and the possibility of reconciliation of the perfections of the Divine Author with the blunders and immoralities of the work.[7]

She began to study heretical books, including the works of Robertson, Stopford-Brooke, Stanley Greg, Matthew Arnold, Liddon, Mansel and others. She found relief from mental strain in practising social welfare work, nursing the sick and helping the poor. She then learnt many truths regarding agricultural labour and she studied the activities of Agricultural Labour Unions which at that time, incidentally, were opposed by the farmers who gave no work to a Union man. She then came across the Rev. Charles Voysey, the atheist preacher. She began to emphasise the humanity of Christ at the expense of the divinity of Christ. She was then faced with the problem that if she gave up belief in Christ as God, she must give up Christianity as a creed. She had a brief correspondence with Pusey but was not convinced. By 1872 she came across Mr. and Mrs. Scott who made their home a centre of heretical thought and Annie Besant's Free Thought essay was written for Thomas Scott. As she remarks picturesquely, it was on the deity of Jesus of Nazareth, and was written, as she herself expressed it, by the wife of a beneficed clergyman. She thereafter declined to attend the Holy Communion and, when asked, she boldly stated that she was unable to take part honestly in the profession of faith required on the part of the communicant. It was in 1872, that tormented as she was by her doubts and living an unhappy domestic life, she nevertheless engaged herself in strenuous nursing activity during a severe epidemic of typhoid fever in the village of Sibsey. And it was at that time that she paid a momentous visit to the Sibsey church. The idea suddenly struck her that she must preach and she felt that she could speak if she had the chance.

In her own words, "So, locked alone in the great, silent church, whither I had gone to practise some organ exercises, I ascended the pulpit steps and delivered my first lecture on the inspiration of the Bible. I shall never forget the feeling of power and delight—but especially of power—that came upon me as I sent my voice ringing down the aisles, and the passion in me broke into balanced sentences and never paused for musical cadence or for rhythmical expression." She then knew that the gift of speech was hers—a gift that was to stand her in such good stead throughout her later life.

By 1873 the marriage tie was broken. The alternative was given by her husband: conformity or exclusion from home; in her language, hypocrisy or expulsion. Her mother was heartbroken but realised the gravity of her problem. But, as she said, she did not know then how cruel man and woman could be, how venomous their tongues. For some time, Annie Besant tried fancy needle work and she became a governess. Her children fell ill and her mother and herself set up house together. Her mother also then fell ill and her case became hopeless. Before she died, she longed for Holy Communion but she declined to take it unless her daughter Annie Besant also took it. Many clergymen would not allow Mrs Besant to communicate but Dean Stanley whom she approached permitted her to do so and she took the Sacrament for the last time. In a discussion that took place at that time between Dean Stanley and Annie Besant, he said that conduct was far more important than theory and that he regarded all as Christians who recognised and tried to follow the moral law of Christ. Her mother died and Annie Besant, in her desperate need, engaged herself in writing pamphlets for Scott. The few guineas that she earned were very valuable. Scott's library was at her service and Annie Besant was feeling her way to the next stage in her life. She attended the sermons of Moncur D. Conway who directed her to the Hall of Science where Charles Bradlaugh used to speak. He told her, "Bradlaugh is the finest speaker of Saxon English except perhaps John Bright, and his power over a crowd is something marvellous. Whether you agree with him or not you should bear him."

It was in 1874 that Mrs. Besant joined the *Free Thought, Her Society* and attended a lecture of Charles Bradlaugh. The description of the experience is arresting:

"The grave quiet, stern, strong face, the massive head, the keen eyes, the magnificent breadth and height of forehead; was this the man I had heard described as a blatant agitator, an ignorant demagogue?

"He began quietly and simply, tracing out the resemblances between the Krishna and the Christ myths, and as he went from point to point his voice grew in force and resonance, till it rang round the hall like a trumpet. Familiar with the subject I could test the value of his treatment of it and saw that his knowledge was as sound as his language was splendid. Eloquence, fire, sarcasm, pathos, passion, all in turn were bent against Christian superstition, till the great audience, carried away by the torrent of the orator's force, hung silent, breathing soft, as he went on, till the silence that followed a magnificent peroration broke the spell, and a hurricane of cheers relieved the tension."

As Annie Besant has often stated, from the first meeting in the Hall of Science dated a friendship that lasted unbroken till death severed the earthly bond which to her stretched through Death's gateway and linked them together. Bradlaugh and she became intimate friends. He was a wonderful man and he insisted:

"You should never say you have an opinion on a subject until you have tried to study the strongest things said against the view to which you are inclined." "You must not think you know a subject until you are acquainted with all that the best minds have said about it." "No steady work can be done in public unless the worker studies at home far more than he talks outside." "Be your own harshest judge, listen to your own speech and criticise it; read abuse of yourself and see what grains of truth are in it."[10]

Strengthened by this friendship and fortified by Bradlaugh's example, Mrs. Besant progressed from stage to stage; and in a chapter in her Autobiography entitled "Atheism as I knew and Taught It", she says:

"Proceeding to search whether any idea of God was attainable, I came to the conclusion that evidence of the existence of a conscious power was lacking and that the ordinary proofs offered were inconclusive; that we could grasp phenomena and no more.

"The Atheist neither affirms nor denies the possibility of phenomena differing from those recognised by human experience... As his knowledge of the universe is extremely limited and very imperfect, the Atheist declines either to deny or to affirm anything with regard to modes of existence of which he knows nothing. Further, he refuses to believe anything concerning that of which he knows nothing and affirms that that which can never be the subject of knowledge, ought never to be the object of belief."

She adds: "For many of us evidence must precede belief. I would gladly believe in a happy immortality for all, as I would gladly believe that all misery and crime and poverty will disappear in 1885—if I could. But I am unable to believe an improbable proposition unless convincing evidence is brought in support of it. Immortality is most improbable; no evidence is brought forward in its favour. I cannot believe only because I wish."[11]

Holding these views, it must, at the same time, be realised, she was filled with a passionate desire for the bettering of the world and the elevation of humanity. To her, therefore, a lofty system of ethics is of much greater importance than a logical, intellectual conception of the universe. In her own language, "...the total loss of faith in a righteous God only made me more strenuously assertive of the binding nature of duty and the overwhelming importance of conduct". It seemed to her, as she asserts in her book *The True Basis of Morality:*

"To the forwarding of this moral growth of man, two things seemed to me necessary—an ideal which should stir the emotions and impel to action and a clear understanding of the sources of evil and of the methods by which they might be drained into the drawing of the first. I threw all the passion of my nature striving to paint the ideal in colours which should enthral and fascinate, so that love and desire to realise might stir man to effort. If 'morality touched by emotion' be religion, then truly was I the most religious of Atheists, finding in this dwelling on and glorifying of the ideal full satisfaction for the loftiest emotions."[12]

She had an instinctive certainty of man's innate grandeur that governed all her thought, inconsistent as that certainty was with her belief in his purely animal ancestry. The *Gospel of Atheism* was published in 1876 and the *Gospel of Christianity and Free Thought* about the same time. She began to study Darwin, Herbert Spencer and Huxley, Buchner and Haeckel, and began to see in the evaluation of the social instinct the explanation of the growth of conscience and of the strengthening of man's mental and moral nature. In her own words:

"The Atheist desires personal perfection not only for his joy in it as beautiful in itself, but because science has taught him the unity of the race, and he knows that each fresh conquest of his over the baser parts of his nature and each strengthening of the higher is a gain for all and not for himself alone."[13]

This was her creed of ethics and conduct from 1874 to 1886. Later on she was to assert that the chief debt of gratitude she owed to Free Thought was that it left her mind open to new truths and encouraged the most unshrinking questions. She worked side by side with Charles Bradlaugh; and, gradually but definitely, great public antagonism grew up against them. Many persons believed that atheism implied degraded morality. Accusations were made against Mrs. Besant of personal immorality and, as she says, her extreme political views combined with her atheism engendered a pervasive feeling of hatred. She began to speak in public on the land question, the cost to the country of the Royal family, the obstructive power of the House of Lords; and she advocated complete Home Rule in colonial countries, and also spoke against the Government policy in Ireland, in Transvaal, in Burma, in Egypt and, later on, in India. Bradlaugh and she worked against capital punishment and against flogging and they demanded a national system of education and of public libraries. She has described her joint work with Bradlaugh in these words:

"For many years he was wont to come to my house in the morning after the hours always set aside by him for receiving poor men who wanted advice on legal and other matters—for he was a veritable poor man's lawyer, always ready to help and counsel—and, bringing his books and papers he would sit writing, hour after hour, I equally busy with my own work, now and then, perhaps, exchanging a word, breaking off just for lunch and dinner, and working on again in the evening till about ten o'clock—he always went early to bed when at home he would take himself off again to his lodging about three quarters of a mile away."[14]

Mrs. Besant accepted a place on the staff of the *National Reformer* newspaper and was paid a weekly salary of a guinea. She began to write under the *non-de-plume* "Ajax". She says that her first attempt at public speaking was in an informal debate at a garden party. Her first lecture was delivered at the Co-operative Institute in 1874. She selected as the theme "The Political Status of Women". Though she was nervous to start with, but the moment she was on her feet and looked at the faces before her, she felt no tremor or nervousness, and as she heard her own voice ring out, she was conscious of power and pleasure, not fear. Once on her feet, she felt perfectly at ease and was in her own language, "ruler of the crowd, master of myself".

It was about this time that Bradlaugh, with the support of John Stuart Mill, stood for Parliament, his constituency being Northampton. Mrs. Besant took part in the electioneering. The election gave her the first experience of rioting. Bradlaugh was accused of all un-nameable vices. It was known that he was separated from his wife, and it was alleged that, being an atheist, he was an opponent of marriage and had deserted his wife and children. He could have easily defended himself as it could be proved that his wife had taken to drinking and her habits were the cause of the break-up of the home; but he would say nothing. Public feeling was aroused against him; and, during the election, the crowd charged towards the door of a house where he was sitting. There were, however, many defenders of Bradlaugh, and it is remarkable that although he was often attacked, he equally often rescued those who had maligned him from being manhandled by his followers.

From 1875 Mrs. Besant started her career as lecturer on Free Thought and undertook lecture tours. She says that lecturing acted as a tonic. She consulted a doctor on the possibility of her being able to stand on a platform and speak. The answer was, "It will cure you or kill you", and she says that it cured her and she grew strong and vigorous. During her tours, she was subjected to many attacks. She was accused of being in favour of the abolition of marriage. She was also falsely denounced as the author of the book *The Bible of Secularists*. The fact was that it had been written by a Doctor of Medicine some years before, and sent to the *National Reformer* for review, as to other journals, in the ordinary course of business. It consisted of three parts—the first advocated from the standpoint of medical science, what is roughly known as "Free Love"; the second was entirely medical; the third consisted of a clear and able exposition of the law of population as laid down by the Rev. Mr. Malthus, and following the lines of John Stuart Mill, the book insisted that it was the duty of married persons voluntarily to limit their families within their means of subsistence. Bradlaugh in reviewing the book, said that it was written "with honest and pure intent and purpose", and recommended to working men its exposition of the law of population. His enemies took hold of this recommendation, declared that he shared the author's views on the impermanence of the marriage tie, and despite his reiterated contradictions, they used extracts against marriage from the book as containing his views.[15]

The National Secular Society was founded about this time. Bradlaugh was its President and Mrs. Besant was the Vice-President. She continued as such until she joined the Theosophical Society.

Mrs. Besant came under the compelling influence of Blavatsky and joined the Theosophical Society. She gave up working for the doctrines for which she had sacrificed so much. She accepted the doctrine of Madame Blavatsky that theosophists should exercise self-restraint and use no other means for checking population. As a result, as she says, though she shrank from the decision, she refused to reprint *The Law of Population* or to sell the copyright. This gave great pain to many friends who had stood by her so valiantly, as she exclaimed, "will it always be, I wonder, in man's climbing upward, that every step must be set on his own heart and on the hearts of those he loves?" About this time, Mrs. Besant had suffered from a long and dangerous illness but she returned to her work with zeal both with reference to her *Free Thought* publications and the struggle against the Disraeli Government's policies. She wrote in 1878, a book which ought to be better known than it is, entitled *England, India and Afghanistan.* In it she anticipated her later pleas for Home Rule for India and against Imperialist expansion.

It was significant that long before Mrs. Besant became a theosophist, she was fascinated by India. As early as 1885, she wrote some tales for children in the "Young Folks Library", and in the first story, "Ganga and the River Maid", she gave a description of India. At the time Mrs. Besant came into the theosophical world, there were only a few books of theosophical literature such as *Isis Unveiled, The Secret Doctrine* and *Esoteric Buddhism.* But as Jinarajadasa says: "When Dr. Besant became a theosophist, she did something which nobody else could have done. She took the Truths of Theosophy from these books and presented them in public addresses, full of idealism and beauty. That is the remarkable gift which she had; she gave beautiful settings for all these wonderful truths. We find her always giving to Theosophical work on an aspect which never existed before. I think particularly of one lecture of hers, which I did not hear because I was in America at that time. It was a lecture called 'The Spiritual Life for the Man of the World', given at City Temple; it is exquisite in form, and there could not be a more artistic presentation of a great topic. One can take the lecture, analyse it section by section, and see how the whole subject is presented with a perfect symmetry. She had also given a set of lectures entitled 'The Outer Court' and then later 'The Changing World'. All these have an artistic structure, and the second work plans for an ideal united world.

"I think Dr. Besant can be described more as an artist than as a lecturer of the professional type, because she does not survey any subject in its entirety, but gives a definite presentation of it from some one angle which she selects, as a painter does when painting a landscape. While she gives Wisdom, she gives it in an exquisite setting. Of her many works, perhaps the one that will last longest is that most fascinating book, *A Study in Consciousness*, for, in it there is a depth of philosophical presentation that will fascinate many a generation to come."[16]

Bishop Leadbeater, according to C. Jinarajadasa (who succeeded Mrs. Besant as President of the Theosophical Society), developed the power of clairvoyance slowly; but according to him, Mrs. Besant developed it almost overnight in 1896.

From the moment that Mrs. Besant landed in India, she began strenuously to work for the all-round regeneration of the country. To quote C. Jinarajadasa again, "Now we come to another facet, the Regeneration of India. The moment Mrs. Besant landed in India in 1893, she set that work before herself. She lectured on the best that Win Hinduism, she preached a new type of education, and she urged Theosophical Lodges to start new schools, and to keep the keynote of Indian ideals when teaching Hindu boys and girls. Before she landed in 1893 she already spoke of India as her Motherland. That phrase is in a letter she wrote in 1892."[17]

She was one of the chief inspirers of the Women's Indian Association and its first President. This association through the influence of Mrs. Besant and her colleagues spread out all over India, from Cape Comorin to the Himalayas, from Peshawar to Sylhet in Assam. The whole country was covered with a network of this women's organisation which worked for furthering the progress of women in education, industry, politics, as well as other aspects of women's uplift. Thus the women were banded together for the service of their country, so much so that when the Southborough Commission visited India, Mrs. Besant and Mrs. Naidu and the members of the Women's Indian Association (W.I.A.) waited on Montagu, the then Secretary of State for India, and presented a memorandum which stated that Indian women also should be made eligible for the franchise on the same terms as men. Mrs. Cousins took a leading part in this movement as also Mrs. C. Jinarajadasa, the then Secretary of the W.I.A. Both these ladies were theosophists and accompanied Annie Besant from their native country to India. Mrs. Cousins and her husband, Dr. Cousins, had adopted this land as their own and had been rendering yeoman service to the Indian people in all walks of life.

It must not be forgotten that Annie Besant was the prime factor in influencing men and women of other lands in favour of Hindu thought and culture at a time when India's own sons and daughters were becoming strangers to their own religion and culture and were beginning to think that their ancestors had been barbarians and forgetting to learn even their own mother-tongue, much less the language of the Vedas and the Upanishads. It was Annie Besant who not only translated the Gita into simple and pure English, but also compiled small booklets containing tales of Aryan greatness for the use of little boys and girls. She also produced a universal Textbook of Religion and Morals which is a compendium of the basic truths of all great religions. Through her efforts mainly, as well as those of Swami Vivekananda, the Western world came to know something of India's spiritual treasures which were engulfed in centuries of darkness and oblivion. She was the one who brought the light and illumined the hidden treasures of Indian thought.

One of the first tasks to which Mrs. Besant committed herself in India was the founding of the Central Hindu College at Benares in 1898 with the co-operation and support of many Indian colleagues, only some of whom were theosophists. She was convinced that the direction of education in India must be in the hands of Indians and that they should shoulder the monetary and administrative responsibilities of evolving educational institutions. She insisted that the guiding spirit of Indian education should be patriotic in outlook, that it should not be divorced from the essence of religion and that it should take the utmost possible advantage of western science and technology.

In the beginning of her life in India, Dr. Besant declined to enter the political field. She recognised that the awakening of India's national life should be based on religion and spirituality. Therefore, she worked incessantly to present the truths of Hindu religion in modem garb in order to carry conviction to the minds of people trained to examine all problems from the intellectual point of view. In order to train the younger generation to feel proud of the glorious heritage of their ancestors, to fill them with a lofty sense of patriotism and love for their Motherland, and to inspire them to serve and sacrifice for the restoring of her greatness and glory, she became the founder and pioneer of Indian National Education. As India was peopled by the followers of all the great religions of the world, she preached the Brotherhood of Religions and the common source from which they have originated and the common basis underlying them all. She aimed at making the younger generation not only proud of their country and its past achievements, but also to lead well-disciplined lives, to build a strong noble character, and to devote their talents to the service of their country. She found that there were fissiparous tendencies endangering the rise of a united nation, and, like a far-sighted statesman, she recognised the necessity for bridging the widening gulf between the Hindu and the Mussalman, which was accentuated by the communal electorates created during Lord Minto's viceroyalty; and she emphasised the recognition of the Brotherhood of Religions as a great solvent for the discord and differences in Indian social and political life. Pandit Madan Mohan Malaviya was insistent upon securing a Charter for a Hindu University, though he did not have a constituent College which would be a preliminary requisite for the founding of a Hindu University. As the British Government were not likely to agree to the chartering of two Universities and as she considered that one denomination of Hindu University was better than none, she gave her Central Hindu College as the Constituent College of the University. The foundation of a Hindu University at Benares gave an impetus to the foundation of a Muslim University at Aligarh. As Dr. Besant feared, these two denominational Universities could not but breed separative tendencies. In this as in many of her actions, Dr. Besant showed her statesmanship, and played the part of a practical idealist, though her views did not often prevail against the lack of political insight and the obscurantism of some of the other national workers. After placing her religious and educational work on a sound basis, Dr. Besant turned her attention to the reform of social abuses, especially infant marriage and the extortion of bride-monies.

The First World War began in 1914, and Dr. Besant, though proclaiming that the allied nations would achieve success even though external indications at that time were to the contrary, stressed that the price of that victory should be the achievement of India's freedom if the results were to be permanent and avoid future wars, and stressed the right of India to work for her freedom even though England was in difficulties. In December 1915, when the session of the Indian National Congress was held in Bombay, she called a meeting of the Congress leaders and announced her determination to unfurl the flag of "Indian Home Rule". The leaders were startled, fearing that the association of that ominous name in connection with the fight for Irish freedom would bring a similar trend of events in the smooth and placid waters in the Indian political stream and eyed her advent as that of a stormy petrel portending unwelcome disturbances within the near political horizon. They said they would take the movement for Indian freedom into their own hands and guide it in their own ways. She gave them time till the 1st October to follow their course of action, after which if they did not act, she would start her Home Rule movement.

September 3, 1916, came but no action on the part of the Congress leaders was visible. On that day she founded the Home Rule League for India. Indian political activities entered a new phase of vibrant life under her leadership. At the Lucknow Congress of 1916, the Hindu and Muslim leaders were led to unite, and the Lucknow pact, though disfigured by communal electorates and weightages for minorities, was entered into for the attainment of India's freedom; as Dr. Besant foresaw, without unity freedom could not be attained. Dr. Besant's political activities during her Home Rule campaign brought repeated forfeitures of security by her paper *New India,* accompanied by battles in the law courts waged by her in defence of her cause which was widely popularised thereby. Her internment with G.S. Arundale and B.P. Wadia raised tremendous enthusiasm throughout the country for the cause of Home Rule, at the height of which she was made President of the Congress session in 1917 in Calcutta.

Mrs. Besant's method to attain freedom for India was by way of intense and well-organised constitutional agitation, educating the masses politically, and seeking their enthusiastic support for the cause, and presenting in the name of the united Indian Nation a Swaraj constitution suited to Indian genius and Indian conditions which the sanction of the Indian people behind it would compel the British Parliament to pass. At this stage Gandhi entered the political field, assuring that Swaraj would be won with his weapon of "Satyagraha" or "Civil Disobedience" of laws in order to hamper and bring to a deadlock the functioning of Government activities.

Dr. Besant was strongly against civil disobedience as she felt it would lead to disturbances and bloodshed, as the masses would get out of hand; and she strongly advocated the British connection and India's place as an equal member in the British Commonwealth leading ultimately to a common Government Federation for the whole world. Gandhi's method was more spectacular and, caught the popular fancy, and Mrs. Besant's practical idealism was set aside.

The world wars and their several crisis, the revolt threatened by the Indian Military and Naval forces, the growth of world opinion in favour of the enfranchisement of subject peoples, the hardly suppressed irritation of the United States over the failure of England to liberate India and, above all, the devoted self-abnegation of large masses of the Indian people whom Gandhiji's life and example galvanised into unity of thought and action in spite of obstacles and suffering—these factors combined with the belated but most remarkable and creditable second thoughts and the basic political maturity of Great Britain brought about a compromise culminating in India's political freedom as a member of the British Commonwealth. It was this ideal for which Dr. Besant strove, suffered and battled. She did not live to see this climax as she rapidly lost her health after her disappointments over the Round Table Conferences of 1930, 1931 and 1932. She passed away in 1933.

REFERENCES

1. Aiyar, C.P. Ramaswami, *Annie Besant* (New Delhi: Publications Division, Ministry of Information and Broadcasting, Government of India), 1963., p. 1.
2. *Ibid.*, p. 2.
3. Nethercot, Arthur H., *The Last Four Lives of Annie Besant* (London: Rupert Hart Davis), 1963, p. 29
4. Aiyar, C.P. Ramaswami, *Annie Besant, op. cit.*, p. 4.
5. *Ibid.*, p. 5.
6. Besant, Annie, *Autobiography* (London: J.M. Dant), 1893, p. 93.
7. Aiyar, C.P. Ramaswami, *Annie Besant, op. cit.*, p. 6.
8. *Ibid.*, p. 7.
9. Nethercot, Arthur H., *The Last Four Lives of Annie Besant, op. cit.*, pp. 39-40.

10. Besant, Annie, Autobiography, op. cit., pp. 74-75.
11. *Ibid.*, p. 11.
12. Besant, Annie, *The True Basis of Morality* (Adyar : The Theosophical Publishing House), 1950, p. 20.
13. Besant, Annie, *Autobiography, op. cit.*, p. 79.
14. Nethercot, Arthur H., *The Last Four Lives of Annie Besant, op. cit.*, p. 75.
15. Aiyar, C.P. Ramaswami, *Annie Besant, op. cit.*, p. 13.
16. *Ibid.*, p. 48.
17. *Ibid.*

2

IDEOLOGY AND PERCEPTION

Greatly influenced by the philosophy of radicalism advocated by Roberts of Manchester and her contacts with free-thinkers and socialists like George Bernard Shaw and Bradlaugh, Annie Besant defied British system of Government in India, glorified Indian culture and heritage, demanded self-government for India, and criticised the violent means used and advocated by extremists. Her speeches and writings had the glaring elements of force, strength, determination and fearlessness. She emphasised the well-known poetical dictum of James Russell Lowell: 'They are slaves who fear to speak for the fallen and the weak.'

Here was a novel method of propagating the principles and ideology of the Home Rule Movement. It was mainly through the media of discussions amongst various groups, organization of a series of lectures in district headquarters, launching of fund-raising campaigns amongst the rich philanthropists and common people, publication of literature—books, newspapers and pamphlets with main emphasis on swarajya, conduct of personal tours to various, regions of India, by seeking the support and sympathy of universities—authorities, teachers and students, and last but not the least, to send deputations to England in order to mobilize British public opinion.

Some of the speeches and articles have been recorded by the government in the official records available in the National Archives of India. The file is fairly comprehensive in contents and contains 277 pages. Besides the extracts deal with national songs, articles by Mrs. Besant, A.O. Hume and others.[1]

Besant believed that by understanding the past Indian traditions, the future of India could be shaped in a proper manner. She, therefore, shaped her programme in such a way which had a social and religious background based on the preservation of India's ancient traditions and culture. She sought to adopt western democratic system to Indian conditions.

The critical assessment of her writings did not dampen her interest, courage and strength. Time and again she talked of India's heritage, the urge for awakening, and efforts for the development of educational system in order to prepare the people for Home Rule Movement.[2] 'For many long years past I have urged on you, and on those like you in all parts of India, the necessity of a spiritual awakening before the awakening of a material prosperity became possible. You know that, during many years past, since the Theosophical Society was founded on the shores, the importance of religion, the necessity of spiritual knowledge, has been constantly insisted upon, has been constantly urged; and in doing this those who brought the renewal of the message were only treading in the footsteps of their far-off predecessors, who have ever declared that from the spirit come forth all things that exist, and that without the life of the spirit not even animal, vegetable or mineral life were possible. That profound truth in the ancient philosophy of India is the only foundation for progress of every kind: one spirit and one only; one life and none other; every form from the one living essence, every being rooted in the everlasting one.

Besant was neither a moderate nor an extremist, but a radical, a bridge between the two. Like a believer in radical change of a constitutional nature, she proposed and indeed forcefully about the widespread governmental changes and reforms at the earliest. She termed moderates as 'yesterdays' and extremists as 'tomorrows'.[3] She may be called a moderate extremist. Undoubtedly she did not advocate or support piecemeal methods and pledged to the oft-repeated demands put forward by the leadership of the Indian National Congress in their annual sessions. And for the achievement of her aim she was to go to a point which touched the extreme.[4]

There were mainly four different ideologies prevailing in Indian politics. From 1907 to 1915 Liberals or Moderates led the national movement. They tried to apply to India the liberal doctrines of the West in such spheres as civil rights, judicial reforms, and constitutional agitation. They realized that in the fields of social and economic matters, the problems of India were unique and that they could not, therefore, find their solution by applying mechanically Western ideas. Thus, the political ideology of the Moderates was partly based on Western ideas, "which they found to be of advantage to India, and partly on their understanding of the Indian situation and their feeling that some of the Western ideas could be modified to suit Indian conditions".[5]

The next phase of the Indian politics was controlled by a group of leaders who were often termed as Extremists. They were radicals in their political programmes. Reference to the past glories of India formed part of their programmes. Religion played a significant role in their political campaign. There was obvious connection between their religious and political movements. The spiritualization of politics was associated with Aurobindo Ghose, and the other was the effort of a leader like B.G. Tilak to arouse enthusiasm by appealing to the religious traditions of the people. The first tried to offer "a theological explanation for political events and pictured India as a Mother who should be worshipped", the other used religious festivals for arousing enthusiasm in political movements and for organizing political agitation on the model of those festivals. The Extremists held that social reforms should not be undertaken prior to political freedom. They were aware of the basic economic conflict between the forces of Indian nationalism and British vested interests. Their political activities comprised economic weapons such as Swadeshi and boycott. They, therefore, promoted vigorous economic nationalism along with political radicalism. A vital feature of the political ideology of the Extremists was their emphasis on the collective freedom of the nation, as distinct from, if not opposed to, the freedom of the individual. In this respect their views were in glaring contrast with those of the Moderates who always emphasized the civil and other rights of individuals.[6]

While assessing the political role of the Extremists, it would be seen thus that their ideals did not form a consistent whole and that they were never crystallized into "a systematic political thought".

Often they varied from one individual to another, and in the case of some individuals from one period of time to another. It is also significant that the Extremist leaders were indifferent to the fundamental questions relating to the structure of the State and of the government, and in the individual's relations with them. They were mainly interested in the immediate question of evolving a strategy and tactics for conducting political agitation with the object of freeing their country from foreign domination. Their political activities and ideas left a deep influence on the political philosophy of the Indian National Movement.

Although many Muslims were influenced by the same set of ideas which influenced other Indians, a large section among them differed from the Hindus on political questions. Two of the reasons given by many of their leaders for this attitude were their adherence to Islam, and the existence of Hindu communalism in the country. The separatist tendencies among two great communities of India were generally held to be due to the "Divide and Rule" policy of the British Government. However, a good number of Muslims cherished the belief that they were a separate political body. Prominent leaders who made a mark on Muslim politics during our period were Sir Syed Ahmed Khan, Ameer Ali, Badruddin Tyabji, R.M. Sayani, Abul Kalam Azad and Mohammed Ali Jinnah. According to Karunakaran, "main trends of thinking among the Indian Muslims were non-involvement in political activities, and concentration on matters connected with their social backwardness, education, and economic progress, cooperation with the rest of the people in their constitutional agitation for the extension of civil rights and political progress, indifference towards political views of Extremist leaders, separate electorate, and weightage for the Muslims and formation of the Muslim League to protect their political interests and developing mood of close cooperation with non-Muslims in the political sphere."[7]

The nature and character of the British administration in India were determined both by political ideas prevalent in the United Kingdom and by the structure of the Government of India. The attitude of Englishmen in India was thus described: "The missionaries of English civilization in India stood openly for a policy of assimilation.

Britain was to stamp her image upon India. The physical and mental distance separating East and West was to be annihilated by the discoveries of science, by commercial intercourse, and by transplanting the genius of English laws and English education. It was the attitude of English liberalism in its clear, untroubled dawn, and its most representative figure in both England and India was Macaulay."[8]

The government ideology was best explained by *The Times,* London, which wrote :

> "The educated classes may find fault with their exclusion from political rights. Political privileges they can obtain in the degree in which they prove themselves deserving of them. But it was by force that India was won, and it is by force that India must be governed in whatever hands the government of the country may be vested. If we were to withdraw, it would be in favour of not of the most fluent tongue or of the most ready pen, but the strongest arm and the sharpest sword."[9]

Annie Besant refused to subscribe to the view that India was conquered by sword, and was held by sword. She believed that India was conquered by her own sons siding with Britain against local hereditary enemies, State against State and British astuteness used Indians to subdue Indians, and by playing off local jealousies against each other, she conquered each State in turn.

World Empire Based on Brotherhood

After having firstly studied various political ideologies prevalent in India, it would now be proper to analyse the influences that moulded Annie Besant's political ideology in Indian context.

Annie Besant piloted and participated in many political activities in the seventies and eighties of nineteenth century in England. Especially her work for the Parliamentary elections of England had given her a great insight into various aspects of practical politics. But the situation in India was different. Her views had undergone a vital change. Some of the main factors which brought about the change were: (a) Spirituality of India and her growing belief in greatness of the Indian thought; (b) Her new-found idealism in theosophy; (c) Indo-British friendship. It was mainly the spirituality of India which had the greatest influence on her political ideology.

She believed with Havell that it would not profit India to gain the whole world and lose her own soul.[10] The disappearance of Indian ideals, she thought, was as sudden as it was disastrous; invasions and even the establishment of a foreign Empire and foreign Kingdoms previous to the invasion and triumph of the East India Company in 1757 had not touched the soul or spirit of India. She pointed out that India had been invaded, but she had assimilated the invaders, and had enriched her own culture by theirs. Portions of her land had been conquered and occupied, but she had turned the conquerors into Indians. The East India Company, she explained, not only drained her of her accumulated wealth and reduced her to poverty but despised her learning and her arts, crushed her with ignorance, and "filled the palaces of her princes with Brummagen imitations and glass-legged sofas and chairs". It destroyed her self-respect and jeered at her religion and tradition. It inculcated in them English attitudes and ideals through the study of English literature and history. Indians longed to find Liberty among the blessings of British rule, but they did not find it. But after studying Indian history and assimilating its lessons, she said:

> "We have resolved to revive the Ancient Ideals of Indian Education and Indian Culture, to make Indian Ideals the basis of Indian Civilization, renouncing the hybrid and sterile ideas of anglicised-Indianism."

Mrs. Besant gave due importance to peculiar characteristics of each nation. She wanted every nation to develop its own ideology on its own lines. She held that every nation had a right to evolve its own ideology and to carve out its future which, to be successful, should be in consonance with its past. She, therefore, wrote that those who sought to build for India foundations of a happy future might well wish to "disentangle the principle of the fourfold order (the various *Ashramas*) from the rubbish which overlay that in modern times, so that the nation might have the benefit of the national tradition, deep rooted in its nature, and thereby evolve in sober and orderly fashion, and avoid the social conflict which threaten Western civilizations".[11]

She believed that the future was not with things material or mundane; it was with the things of soul and spirit. The body dies, but the soul is immortal: civilizations rise and fall, but the spirit of man lives for ever. She held that India was the land of spirituality, and a great civilization. She said:

"If religion perishes here, it will perish everywhere, and in India's hand is laid the sacred charge of keeping alight the torch of spirit amid the fogs and storms of increasing materialism. If that torch drops from her hands, its flame will be trampled out by the feet of the hurrying multitudes, eager for worldly good, and India bereft of spirituality will have no future but will pass on into darkness as Greece and Rome have passed."[12]

In her Anniversary Address in February 1900 she expressed that with the revival of spirituality alone, India could regain her greatness, and behind it would follow other things, intellectual strength, material wealth, and all the lower things, that enter into the growth of national life.[13]

India's ancient character, she believed, was preeminently spiritual, and the social polity that was its natural expression was shaped to give effect to spiritual ideas and to subordinate the lower nature to the higher, so that the nation might be the "school of souls", and growth and development of the soul might be on every hand aided and encouraged. From that past it had resulted in the present, that India, even in her present low state, despite the loss of spiritual life and the almost extinction of spiritual fire, has remained "the one country where spirituality still hovered in the very atmosphere and where external surroundings helped the soul to rise instead of fettering it to earth".[14]

She added: "Out of the spirituality grow reverence to parents, teachers and elders and reverence to the gods translated itself in the family and social life into reverence for the parents, who gave and nourished the physical life; for the teacher who gave and nourished the inner life—the second birth; for the aged, whose ripe wisdom served as guide and who handed on the ancient traditions.

"From reverence sprang courtesy, respecting others and self-respecting; then came hospitality, a duty ungrudgingly done; and its sister charity, so that none might starve while others had enough and none of the nation's children must go unfed. That reverence, courtesy, hospitality, charity—these were the social virtues of the Aryan householder that rendered him so gracious a type. Rigid fidelity to the pledged words, undeviating accuracy, these were taught by sacred precept and lofty examples. Courage and fearlessness were other distinguishing characteristics of the Aryan type.[15]

In a lecture delivered in 1895, Annie Besant expressed the view that if India was again to hold up her head among nations, India's younger children must begin to lay the foundation in their own life of the Aryan type of character. These virtues, she pointed out, were the most pronounced attributes, and the revival of those among the Aryan youth would presage the rebuilding of the nation. "Character makes destiny, and Indian destiny depends on Indian character".[16] Thus, spirituality of India became the keynote of Annie Besant's political ideology.

The second great influence was her new-found idealism in theosophy. It convinced her that an empire could be founded on brotherhood, on righteousness and on truth, an empire which was based upon peace, justice, within which a new civilization might gradually grow up, which should stand for peace, not war, cooperation, not competition, education, not cramming, comfort, not pauperism. In a lecture that she delivered on "Theosophy and Imperialism" in 1902, she said:

"I believe, thoroughly believe, that at the present time to this British nation the possibility of a world Empire is offered. I believe that in the cycle of evolution and the growth of the peoples, the time has come in the vast world-history where this power of serving the world is offered to the British nation that I believe to be true. I believe it because I am a Theosophist, and have studied history in the light of occultism. How vast a destiny for Britain, how magnificent a possibility for the world, if this nation can rise to the greatness of such a destiny, if this nation can be heroic enough to hold and guide and uplift. For it would mean nothing less than a world-peace, amid which a mighty civilization might grow up greater than the past has seen. It would mean to the world a federation so strong of peace-living nations, that they would be able to impose peace upon the world because none should be strong enough to break it. And if the world is for such world-wide peace, so that the problems may be dealt with which are threatening the present civilization, and the nations may have time to look at home instead of always keeping anxious eyes abroad."[17]

But Mrs. Besant was fully aware of the limitations of the British Imperial polity. She pointed out the defects in the Imperial ideology while ruling a highly civilized nation like India. She explained the requisite characteristics of a sound Imperial ideology for the Empire of her kind. She warned the British nation that had not yet developed that "Imperial insight which ruled a nation on lines suited to the nation that was ruled, instead of on lines suited to the nation which was governing". That was why, she said, blunders were being made in India:

"You are dealing in India with a civilization far older than your own and a civilization suited to the national genius; you have to live among traditions interwoven in the hearts and lives of the people, traditions which it is folly to ignore, which it is madness to outrage and insult."[18]

She pointed out that in dealing with a highly civilized nation the rulers must adapt themselves to the conditions of that nation evolved through ages, and not impose on them conditions alien to their ideals though agreeable to the rulers themselves. She added that the methods of land taxation, economic systems, which were suitable for Great Britain, did not suit the vast Asiatic nation whose traditions and whose customs were utterly different from the rulers .[19]

She remarked that India was not regarded as part of the Empire but as a land of exile. Men did not go to make their home there "to love and sympathise with the people among whom they lived: they went there to make money, longing for the time to return and spend the same at home". She explained that India was not ruled for the prosperity of the people, but for the profit of her conquerors. She added that over 17 million sterling a year were taken from her as

"Home Charges" to be spent in England while English officers in India drew "abnormally high salaries".

Annie Besant believed that there was little effort to understand an "ancient, conservative and aristocratic people", and the real virtues of the Englishman, "his conscientiousness, his diligence, his wish to do justice were masked by a repellent demeanour and a chilly superciliousness of bearing". She added:

"Nor is there anything in this huge bureaucracy to arouse the instinct of loyalty so deepseated in the Indian breast. The crowd of officials veils the Crown, and the Monarch is hidden behind a mass of clerks. The Viceroy with his five years' term appointed in England for political reasons, appears more as the head clerk of a great system of clerks than as a symbol of an Emperor, and he cannot rouse the personal loyalty which in India means power."[20]

She suggested that it would be better to place on the Indian vice-throne a prince of the Royal House, a living representative of the Imperial Crown, who would be assisted by the wisest and best in India, who would rule as well as reign.[21]

She cautioned that if England was to be truly Imperial, "she must think of duty, of industry, of diligence in the discharge of duty, in every rank of life; and the ideal of life must cease to earn money and to live in luxurious idleness". She explained that from the King on his throne to the poorest labourer in the street, the ideal for both should be that of duty and service. She said that unless that was done, the dream of Empire would vanish, and "the opportunity offered to Britain would pass on to some other nation". England's worthiness for the next world Empire, she held, "depended upon her will and capacity to imbibe and practise the noble qualities of head and heart, and not in the pageantry and glitter of the Empire".

Mrs. Besant pleaded with England to adopt the following attitude:

"Let your ambition to be known as the helper of the weak, the protector of the helpless, the one who is ready to stand between the feeble and those who desire to oppress. Let that be your pride, that you hold a shield high, under which the nations of the world may gather, sure of protection, sure of help, sure of justice, sure of sympathy."[22]

Thus, on the basis of her theosophical belief, Mrs. Besant worked for the possibility of establishing a world empire based on the brotherhood of man in which she believed the friendship of India and England was indispensable. Therefore, the next point which impressed upon her political ideology was to strengthen the bonds of friendship between the two countries.

England and India, were bound together, and Annie Besant wanted to strengthen their relations on the basis of equality and brotherhood. She believed that much of the future progress of the world depended on the adjustment of relations between the conquering nation and the subject people. She pointed out that when two nations came into touch with each other, each had something to learn, especially when two such nations as India and England were concerned. India, she said, was enjoying a high state of civilization long before the seed of western civilization was sown. Relations with such a country, she held, must need be complicated and difficult. She said:

"I know of no greater service that can be rendered, either in this land or that, than the service of those who try to understand the question and to draw the nations closer together by wisdom, instead of driving them further apart by ignorance and by prejudice."[23]

Mrs. Besant then explained the ways in which England and India could help each other. She said that with regard to political conditions both the nations had much to learn in mutual understanding and in adaption to the old civilization of India of methods of thought of rule, of social conditions utterly alien from her own conditions. So that changes, if it be wise to introduce, must be brought about "with the greatest care, the greatest delicacy, after the longest and most careful consideration."[24]

Therefore, while emphasizing loyalty to the British throne which stood as the "dignified symbol of a common endeavour", she gave no special place to the English people themselves as a member of the conquering race, or as persons in administrative situation over the people. She fought for equality, inculcating friendliness and modesty among those who ruled, and self-respect among those who were ruled.

A Well-intended Folly

Annie Besant did not overlook the economic aspect in her political ideology. She felt that the prosperity of a country through a beneficial administration revolved round the latter's financial stability, which was to be based on right economic principles. She said that the greatest Indian leaders like Naoroji, Ranade and Gokhale had been able to do much good to the country because economics was "the fulcrum of the lever of their activities". She pointed out that they studied the problems of Indian finance, and had thus rendered most efficient help to the rulers. She felt the need of doing further work along these lines.

Mrs. Besant held that young Indians who were eager to jump into "the roaring sea of Indian politics should have learnt beforehand the economic trick of swimming so essential for their safety, as also to the well-being of those to whom they belonged and to whom they owed an obligation, viz. their own countrymen."[25]

Mrs. Besant pointed out that the British rule in India had "given her famines and poverty by cutting at the very roots of its indigenous economic system, and imposing modern "political economy". She said that "to impose English political economy on India was a folly, well-intended folly, but folly nonetheless,"[26] due to which India faced famines and degradation, and destruction of her arts and manufactures. She remarked that "to sell the corn in the years of plenty and to starve in the years of scarcity might be modern political economy but it was pure idiocy". Likewise the arts of India, she pointed out, depended on the social conditions of the country. The artist in India was not a man "who lived by competition, but he was utterly unable to guard himself against that". As a result, the artist was falling back into the "already well-filled agricultural ranks".

These were the questions, she believed, the rulers had to ask and understand. They had to understand the question of Indian taxation; they had to understand the question of taking away from India "seventeen millions a year to meet English charges". They had to consider "the expenses of their Government in India and the exorbitant salaries that were paid to English officers".[27] They had thus to understand the financial exploitation of the people of India, she said:

"It might be possible to form a band of English Experts, who would be able to make these questions their speciality, and should have weight with the government in India, so that they could advise with wisdom, so that they could point out the most useful path by which improvement could be made."[28]

Mrs. Besant ruled out the possibility of governing a great country like India by a Parliament over here. She believed that it was "too clumsy an instrument for ruling such people." She suggested that "a Council of most thoughtful and wisest men of India should be composed to advise the Viceroy, who should hold his post not as a reward for political service but because he knew and understood India."[29]

A Moderate Extremist

Annie Besant believed that the basis of Indian politics should be national and indigenous. It should be based upon the revival of her ancient literature and language, by awakening the youth of the country to the Indian ideals of life, by building up the nation through encouragement of its national dress, manufactures and arts.

She pointed out in 1895 that reforms to be successful in India must be based upon Indian ideals practised in thought and action.

In order that the actions of Indians may be properly guided and may be inspired to activity, not only does it need to be taught as an ideal from the platform, to be taught as an ideal through the Press, but also in their daily life; that they should make it the subject of deliberation and collective thought, for that thought is after all the greatest force.[30]

While emphasizing the role of the classics in the life of a nation, she pointed out that for hundreds of years in the West the cultivation of classics, Greek and Latin, was regarded as absolutely necessary for what was called education of a gentleman, and those who were ignorant of the classics were regarded as uneducated. She held that the training of intellect gave a certain definite strength and refinement of expression. Similarly, she said that India also had her own ancient language which laid at the root of her vernaculars, a knowledge of which "opened out to her the grandest literature the world had yet produced". The knowledge of that was of vital importance to the development of the nation. She explained:

"If your mental faculties are only directed to the subjects which enable you to keep yourselves alive then you strike at the very root of the development of your nation, and you must sink lower and lower amongst the peoples of the world."[31]

In order that men might be competent to meet the existing needs of the country, she felt it necessary for the people to learn Sanskrit. This, she said, would encourage the people to enrich India's knowledge and experience. She firmly believed that knowledge of past literature and thought would bring out India from her stupor. She said:

"And the pride which grows up in the human heart in feeling itself linked with a mighty past, all that is sympathetic in the past shall become capable of working in a future and impress on that future something of the spiritual greatness which the past has shown."[32]

She believed that the common language was not merely a convenience, but was a tie which bound heart to heart and mind to mind. She said that the idea of having a foreign language as a common language would be laughed at any other country. A common language like Sanskrit which was that language of the ancient books, and mother of vernaculars, she explained, would make men feel "the tie of brotherhood, instead of different races and tend far more to the national feeling".[33]

Annie Besant held that the education of the children should be based on their ancient language and literature because enthusiasm in the young was easily aroused. The young were touched and moved easily by noble ideals. She wrote:

"If you familiarise them with the past history of their own country, if you wake their devotion to their national faith, the time will come when they will turn away from the West to the motherland. And these boys, grown into men, shall be bound with every bond that link the Indian to his home and from such men will come the salvation of India."[34]

As the future was in the hands of the young, she believed that they should be taught Indian history, Indian literature and Indian customs, and that would make a real nation.

Annie Besant held that the entire Indian nation could be built up by the encouragement of national feeling, by maintaining traditional dress, and ways of living. This was the point, she remarked, which went to the very root of Indian revival and was not simply a matter of sentiment. The wearing of Indian dress, she held, was a matter of health, of convenience, and of economy; for the Indian dress suited the Indian climate, not only because it was light, but also because its material could go through water daily, and was more suited to a hot country than the Western dress which was worn unwashed over and over again. There was no reason, she said, no common sense, which should make the Indian lay it aside, when the experience of thousands of years had shown it to be the best kind of dress for India. She pointed out the dangers of Westernization.

"The inner feeling and outer expression often go together and he who Westernises his outer attire is very likely to grow western inside as well, and therefore, instead of strengthening, he really tends to weaken his motherland."[35]

It is likely that such views might be regarded as obsessive if not eccentric.

Mrs. Besant advised the people to lead a national, simple life with all the noble characteristics of ancient times; that their houses should have the old simplicity and not be crowded over with a multiplicity of things of foreign manufacture.

Mrs. Besant emphasized the need to encourage Indian manufacture and art. She explained, how superior were the Indian made goods as compared to the foreign goods. By buying foreign things and by using inferior goods of foreign make, Indians, she felt, were destroying their own beautiful art. She pointed out that Indians were extinguishing the Indian trade by continuing to encourage foreign goods, "to the impoverishment of India and to the throwing of Indians out of jobs". She suggested that public opinion might be formed which would cry "shame" upon a prince or a *raja* "who filled his palace with foreign articles instead of having them produced in his own country so that his wealth would add to the comfort and happiness of the people and strengthen the national prosperity".[36]

Annie Besant's political ideology was thus based upon India's religious, social, economic and national revival. The influences which shaped her political views were the spirituality of India, the Indo-British friendship and her new found idealism in theosophy for the establishment of a Commonwealth of Nations and a World Empire on the basis of equality, love and brotherhood of man. It would thus be seen that her political ideology was both wider and different than the two main political ideologies prevalent in India.

It was wider in the sense that it considered in terms of the salvation of the whole world and not merely of India. Through the revival of India she wanted to elevate the whole world. India was a means but not the end. She thus thought in terms of establishing a Commonwealth of Nations and a World Empire. Like the Moderates she believed in constitutional means. But unlike them she more thoroughly believed in the spirituality of India, and the revival of Indian ideals. She realized that the changed government alone could make possible her works for the uplift of India and its spirituality, which she held was important for the well-being of the whole world. The Moderates did not go so far.

Mrs. Besant differed from the Extremists also. She never wanted to throw the foreign rulers overboard overnight. She wanted India to evolve into a self-governing nation. Like the extremists she believed that the basis of Indian politics should be national and indigenous. She wanted to revive the old customs and traditions in the administration and governance of India. But she liked it to develop through lawful and constitutional means. At times she used such language as came from the Extremists and she also used freely the idiom of the Moderates. But she never advocated violence at any stage. She remained faithful to the Crown thinking it as a symbol of her kind of Empire based upon brotherhood and equality. This was against the ideals of the Extremist section.

For her adherence to constitutional means, Mrs. Besant cannot be considered a Moderate or be labelled an Extremist because she was considered so by the Government. That was one of the reasons why the Government for quite sometime could not make out if her activities were seditious or not. She chided the authorities for misunderstanding, for not being able to really understand her politics.

And when in 1917 her position became clear she was interned by the Government but this too did not earn her the title of an Extremist in Indian politics because she did not use the extremist means. She succeeded for the time being in 1916 in providing a bridge between the Moderates and Extremists. But what was her politics like?

It would thus be seen that she was neither a Moderate nor an Extremist but a Radical. Like radicals she advocated widespread governmental changes and reforms at the earliest opportunities. No wonder, she termed moderates as "yesterdays" and extremists as "tomorrows".[37] One could perhaps legitimately call her a moderate extremist, though it is a wild generalization to classify men and events when they elude classification. She did not agree with piecemeal methods and pledged to the whole demand made by the Congress and like a radical was willing to go to the extreme.[38]

Roberts of Manchester had planted the seed of Radicalism in her. It was nurtured through her contact with Freethinkers and Socialists like Bradlaugh and Bernard Shaw. Indian contacts further added spiritual dimensions to her radicalism in politics. She glorified the past of India, was intolerant of the British methods of government, stood for self-government for India, condemned the violent means used by the Extremists and had implicit faith in the Moderates' constitutional methods. Different strands worked in her and she was radical in politics, eclectic in religion, and patriotic to India's cause while being loyal to the Crown. Vehemence of the Extremist in the use of language and steadying temper of the Moderate combined with the passion of an Orientalist made her unlike a Moderate or an Extremist.

REFERENCES

1. Home Deptt. Pol. A, October 1916, File 36-53.
2. *Speeches and Writings of Annie Besant* (Madras: Natesan), n. d.
3. The *Commonweal*, 5 July, 1918.
4. *Ibid.*
5. *Ibid.*
6. Karunakaran, K.P., *Indian Political Thought 1832-1921* (New Delhi: Gitanjali Publishing House), 1996, p. 12.
7. *Ibid.*, p. 16.

8. Quoted in Eric Stokes, *The English Utilitarian and India* (Oxford: Oxford University Press), 1959, pp. xii-xiv.
9. The Government of India, Documents on the History of Freedom Movement, Vol. II, Appendix C.
10. Indian Ideals, 1930, p. 36.
11. *Ibid.*, p. 38.
12. "The Aryan Type", *India*, Vol. IV, 1895, p. 4.
13. "India's Mission Among Nations", *India*, p. 1.
14. "Anniversary Address", *India*, February 1900, p. 160.
15. "The Aryan Type", *India*, Vol. IV, p. 5.
16. *Ibid.*, p. 6.
17. *Ibid.*, p. 8.
18. *Ibid.*, pp. 197-98.
19. *Ibid.*, p. 183.
20. *Ibid.*
21. *Ibid.*, p. 185.
22. *Ibid.*
23. *Ibid.*, p. 207.
24. "England and India", *India*, October 1903, p. 209.
25. *Ibid.*, p. 210.
26. *New India*, 28 April 1915, p. 8c.
27. "England and India", *India*, Vol. IV, p. 230.
28. *Ibid.*, p. 234.
29. *Ibid.*, p. 235.
30. "The Means of India's Regeneration", *India*, 1895, p. 85.
31. *Ibid.*, p. 93
32. *Ibid.*, p. 94.
33. *Ibid.*, p. 95.
34. *Ibid.*, p. 108.
35. *Ibid.*, p. 114.
36. *Ibid.*, p. 119.
37. *The Commonweal*, 5 July, 1918.
38. *Ibid.*

3

PHILOSOPHY OF THEOSOPHY

Theosophy is Divine Wisdom, and that Wisdom is the Light which lighteth every man who cometh into the world. It belongs to none exclusively; it belongs to each inclusively; the power to receive it is the right to possess it; the fact of possession makes the duty of sharing. Every religion, every philosophy, every science, every activity, draws what it has of truth and beauty from the Divine Wisdom, but cannot claim it as its own against others. Theosophy does not belong to the Theosophical Society; the Theosophical Society belongs to Theosophy.[1]

What is the essence of Theosophy? It is the fact that man, being himself divine can know the Divinity whose life he shares. As an inevitable corollary to this supreme truth comes the fact of the Brotherhood of Man. The divine Life is the spirit in everything that exists, from the atom to the arch-angel; the grain of dust could not be, were God absent from it; the loftiest seraph is but a spark from the eternal Fire, which is God. Sharers in one Life, all form one Brotherhood. The immanence of God, the solidarity of Man, such are the basic truths of Theosophy.[2]

Its secondary teachings are those which are the common teachings of all religions, living or dead; the Unity of God; the triplicity of His nature; the descent of Spirit into matter, and hence the hierarchies of intelligences, whereof humanity is one; the growth of humanity by the unfoldment of consciousness and the evolution of bodies, i.e., reincarnation; the progress of this growth under inviolable law, the law of causality, i.e. *karma*; the environment to this growth, the three worlds, physical, astral, and mental, or earth, the intermediate world and heaven; the existence of divine Teachers, superhuman men.

All religions teach or have taught these, though from time to time one or another of these teachings may temporarily fall into the background; ever they reappear as the doctrine of reincarnation fell out of ecclesiastical Christianity, but is now returning to it, was submerged, but is again emerging.

It is the mission of the Theosophical Society as a whole to spread these truths in every land, though no individual member is bound to accept any one of them. Every member is left absolutely free, to study as he pleases, to accept or to reject; but if the Society, as a collectivity, ceased to accept and to spread them, it would also cease to exist.

This unity of teachings among the world-religions is due to the fact that they are all founded by members of the Brotherhood of divine Teachers, the custodian of the Divine Wisdom of Theosophy. From this Brotherhood come out, from time to time, the Founders of new religions, who ever bring with them the same teachings, but shape the form of those teachings to suit the conditions of the time such as the intellectual stage of the people to whom they come, their type, their needs, their capacities. The essentials are ever the same; the non-essentials vary. This identity is shown in the symbols which appear in all faiths, for symbols form the common language of religions. The circle, the triangle, the cross, the eye, the sun, the star, with many another, ever bear their silent testimony to the fundamental unity of the religions of the world. Understanding this, the Theosophical Society serves every religion within its own domain, and draws them together into a Brotherhood.

In morals, Theosophy builds its teachings on the Unity, seeing in each form the expression of a common life, and therefore the fact that, what injures one injures all. To do evil—i.e., to throw poison into the life-blood of humanity—is a crime against the Unity. Theosophy has no code of morals, being itself the embodiment of the highest morality; it presents to its students the highest moral teachings of all religions, gathering the most fragrant blossoms from the gardens of the world faiths. Its Society has no code, for any code that could be generally imposed would be at the average low level of the day, and the Society seeks to raise its members above the ordinary level by ever presenting to them the highest ideals, and infusing into them the loftiest aspirations. It leaves aside the law of Moses to walk in the spirit of the Buddha, of the Christ. It seeks to evolve the inner law, not to impose an outer. Its method with its least evolved members is not expulsion, but reformation.

The embodiment of the Divine Wisdom in an organisation gives a nucleus from which its life-forces may radiate. A new and strong link is thus made between the spiritual and the material worlds; it is in very truth a Sacrament, "the outward and visible sign of an inward and spiritual grace," a witness of the Life of God in Man."[3]

A Solar System is a group of worlds circling round a central Sun, from which they draw light, life, and energy. On this all Theosophists and non-Theosophists are agreed. But the Theosophist sees much more than this in a Solar System. It is to him a vast Field of Evolution, presided over by a divine Lord, who has created its matter out of the ether of Space, permeating this matter with His Life, organising it into His Body, and from His Heart, the Sun, pouring out the energy which circulates through the System as its life-blood—lifeblood which returns to the Heart when its nutrient properties are exhausted, to be re-charged and sent forth again on its life-sustaining work.

Solar System is, to the Theosophist, not merely a splendid mechanism of physical matter, but the expression of a Life, and the nursery of Lives derived therefrom, instinct in every part with latent or active intelligence, desire and activity. It "exists for the sake of the Self," in order that the germs of Divinity, the embryonic Selves emanated from the supreme Self, may unfold into the likeness of the Parent-God, whose nature they share, being truly "partakers of the divine Nature." Its globes are "man-bearing," and not men alone but also sub-human beings are its inhabitants. In worlds subtler than the physical dwell beings more highly evolved than men, as also beings less evolved; beings clothed in bodies of matter finer than the physical, and therefore, invisible to physical eyes, but nonetheless active and intelligent; beings among whose hosts myriads. of men are found, men who have, for the time, discarded their fleshly raiment, but who, nonetheless, are thinking, loving, active men. And even during life on our physical earth, encased in the garment of the flesh, men are in touch with these other worlds and other world beings, and may be in conscious relation with them, as the Founders, Prophets, Mystics and Seers of all the faiths have witnessed.[4]

The divine Lord manifests Himself in His System in three Aspects, or "Persons," the Creator, the Preserver, the Regenerator; these are the Holy Spirit, Son and Father of the Christian; The Brahma, Vishnu and Shiva of the Hindu; the Chochmah, Binah and Kether of the Hebrew Kabbalist; the Third, Second and First Logos of the Theosophist, who uses the old Greek term, "the Word," for the manifested God.

The matter of the System is built up by the Third Logos, seven types of atoms being formed by Him; aggregations composed of these yield the seven fundamental kinds of matter found in the System, each denser than its predecessor, each kind being correlated with a distinct stage of Consciousness. We call the matter composed of a particular type of atom, a plane, or world, and hence recognise seven such planes in the Solar System ; the two highest are the divine, or super-spiritual planes, the planes of the Logos, and the lower of these two is the birthplace and habitat of the human Self, the Monad, the God in man; the two succeeding are the spiritual planes reaching which man realises himself as divine; the fifth, still densifying, is the intellectual plane; the sixth, the emotional and passionate, the seat of sensations, and desires, is generally called the astral plane; the seventh, the physical plane. The matter of the spiritual planes is correlated with the spiritual stage of Consciousness, and is so subtle and so plastic that it yields to every impulse of the Spirit, and the, sense of separateness is lost in that of unity. The matter of the intellectual plane is correlated with the intellectual stage of Consciousness, with Thought, Cognition, and every change in Thought is accompanied with a vibration of its matter. The W.K. Clifford seems to have recognised "mind-stuff" as a constituent of the cosmos, for, as every force needed its medium, thought, regarded as a force, needed a special kind of matter for its working. The matter of the astral plane is correlated with the desire stage of Consciousness, every change of emotion, passion, desire, sensation being accompanied with; a vibration of its matter. The matter of the physical plane is the coarsest or densest, and is the first to be organised for the active expression of human Consciousness.[5]

These seven kinds of matter, interpenetrating each other—as physical solids, liquids, gases and ethers interpenetrate each other in the objects round us—are not all spread evenly over the whole area occupied by a Solar System, but are partly aggregated into planets, worlds or globes; the three finest kinds of matter do spread over the whole, and are thus common to the system but the four denser kinds compose and surround the globes, and the fields occupied by these are not in mutual touch.

We read in various Scriptures of "Seven Spirits"; Christianity and Muhammadanism have seven Arch-angels; Zoroastrianism has seven Amshaspends; Hebraism has seven Sephiroth; Theosophy calls them the seven Planetary Logos; and they are the Rulers of the Planets Vulcan, Venus, Earth, Jupiter, Saturn, Uranus and Neptune.

Each of these seven Planets is the turning-point in a chain of interlinked worlds, presided over by the Planetary Logos, and each chain is a separate Field of Evolution from its earliest beginnings up to man. There are thus seven such subsidiary Fields of Evolution in a Solar System, and they are, naturally, at different stages of progress. The chain consists of seven globes, of which generally one is of physical and six of finer matter; in our own chain, however, our earth has two sister globes visible to physical sight—Mars and Mercury—and four invisible companions. The wave of evolutionary life, bearing the evolving beings, occupies one globe at a time—with certain special exceptions which need not be mentioned here—passing on to the next in order when the lessons on the earlier have been learned. Thus our humanity has travelled from globe 1, on the mental plane, to globe 2, on the astral; from that to globe 3, Mars, and to globe 4, our Earth; it will pass on to globe 5, Mercury, and from that to globe 6, again on the astral, and thence to globe 7, on the mental. This completes a great evolutionary Round, as it is aptly called.[6]

This huge scheme of evolution cannot be readily grasped by the ignorant, any more than can the corresponding scheme of the astronomer, which deals only with the physical plane. Nor is it necessary that it should be understood by those of small intelligence, since it has no immediate bearing on life. It is interesting only to the man who, desiring to understand, is ready to grapple with the deeper problems of nature, and does not grudge strenuous intellectual exertion.

Man is a spiritual Intelligence, who has taken flesh with the object of gaining experience in worlds below the spiritual, in order that he may be able to master and to rule them, and in later ages take his place in the creative and directing Hierarchies of the universe.

There is a universal law that a Consciousness can only know that which it can reproduce; one Consciousness can know another in proportion as it is able to reproduce within itself the changes in that other. If a man feels pain when another man feels it, happiness when the other feels it, anxiety, confidence, etc., with the other, at once reproducing his moods, that man knows the other. Sympathy—feeling together—is the condition of knowledge, But Consciousness works in bodies; we are clothed, not naked; and these bodies are composed of matter. Consciousness may affect Consciousness, but how can Consciousness affect these bodies?[7]

There is another law, that a change in Consciousness is at once accompanied with a vibration in the matter near it, and each change has its own answering vibration, as a musical sound and a particular length and thickness of string invariably go together. In a Solar System all the separated Consciousnesses are part of the Consciousness of the divine Lord of the system, and all the matter of the system is His Body—"in Him we live and, move and have our being." He has formed this matter and related it to Himself, so that it answers everywhere by innumerable kinds of vibrations to the innumerable changes in His Consciousness, each to each. Over His whole vast kingdom his Consciousness and His matter answer each other in perfect and perpetual harmony and invisible relation.[8]

Man shares with the divine Lord this relation, but in an elementary and feeble way; to the changes in his Consciousness answer vibrations in the matter around him, but this is only perfect and complete, at first, in the super-spiritual worlds, where he exists as an emanation from the Lord; there every vibration of matter is answered by a change in his Consciousness, and he knows that world, his birthplace and his home. But in worlds of matter denser than that lofty region he is as yet a stranger; the vibrations of that denser matter, though all around him, do not affect him, are to him non-existent, as the waves which carry messages by wireless telegraphy do not affect us in this world, and are to our senses non-existent. How then can he grow to the likeness of his divine Parent, to whom every vibration has a message, who can wet up any willed vibration in matter by a change in Consciousness, who is conscious and active at every point of His system ?

The answer comes in the words: Involution and Evolution. He must involve himself in matter, attract to himself an encasement of matter, draw round himself materials from all the worlds—spiritual, intellectual, emotional and physical; this is the involving of Spirit in matter—Involution—sometimes called the descent of Spirit into matter, sometimes the fall of Man. Then, having acquired this encasement, he must slowly try to understand the changes in himself in his own Consciousness—the surging, confusing, bewildering changes that come and go without any will of his, due to the vibrations set up in this material encasement of his by vibrations in the larger world around him, and that force upon his Consciousness unsought changes and moods. He has to disentangle these, to refer them to their proper origins, to learn through these the existence and the details of the surrounding worlds, to organise his own appropriated matter—his bodies—into more and more complex, receptive and discriminative agents, to admit to or shut out from these bodies at will the vibrations that hurtle round them outside, and at last, through them, to impress the changes in his Consciousness on external Nature, and thus to become its Ruler, instead of its slave. This is Evolution, the ascent of the Spirit through matter, its unfolding within a material encasement, drawn from the various worlds which form, its environment, the permeation with its own life of the matter it appropriates, thus rendering it the docile servant of Spirit, and redeeming it from its cruder uses to the service of the liberated Sons of God.[9]

This material encasement, drawn from the different worlds, must be gradually organised, by impacts from without and answers from within, into a "body," or a vehicle of Consciousness. It is organised from below upwards, or from denser to finer, the materials from each world being organised separately, as a means of receiving communications from, and acting upon, its own world. The physical material is first drawn into a fairly compact mass, and the organs which carry on life-processes, and those of the senses, are first slowly evolved; the wonderful and complicated physical body is evolved through millions of years, and is still evolving; it puts man into touch with the physical world around him, which he can see, bear, touch, taste and smell, and in which he can bring about changes by the use of his brain and nerves, directing and controlling his muscles, hands, and feet. This body is not perfect, for there is still much in the physical world around it to which it cannot answer—forms, like atoms, which it cannot see, sounds which it cannot hear, and forces which it cannot perceive, till they have brought about effects by moving large masses of matter big enough for it to see. He has made delicate instruments to help his senses and to increase their perceptive range—telescopes and microscopes to help the eye, microphones to help the ear, galvanometers to find out forces which escape his senses. But presently the evolution of his own body will bring all his physical world within his ken.

Now that the physical body is highly organised, the next finer material, the astral, is being similarly evolved, and is bringing man gradually into touch with the astral the emotional, passional, desire—world around him. Most of the people of the advanced races are becoming slightly conscious of astral impacts, while some are distinguishing them clearly. Premonitions, warnings, conscious touch with the "dead," etc., all are affections of the astral body from the astral world. They are vague and dim because of the poor organisation possessed by this body at present, but those who have forced its evolution are free of the astral world, as everyone is of the physical.

The third state of matter, the mental, is also in course of organisation, and is putting man into touch with the intellectual world around him. As the mental body evolves, the man comes into conscious relation with mental currents, with the minds of others near and distant "living" and "dead".[10]

The spiritual worlds still remain after this for man to conquer, and they have their appropriate body, the "spiritual body", of which S. Paul speaks. This organisation of matter to be the servant of Spirit is the part assigned to man in the great workshop of the worlds, and when the human stage is over there is nothing in the Solar System which he is incapable of knowing and affecting. He came forth from the divine Lord pure indeed, but ignorant and useless outside the subtle region of his birth; he returns, after his long pilgrimage, a wise and strong Son of God, ready to bear his part throughout the ages of the future as a minister of the divine Will in ever-widening fields of service.

The Worlds in which man is evolving as he treads the circle of births and deaths are three: the physical world, the astral or intermediate world, the mental or heavenly world. In these three he lives from birth to death in his waking day-life; in the two latter he lives from birth to death in his sleeping night-life, and for a while after death; into the last he occasionally, but rarely, enters in his sleeping night-life, in high trance, and in it he spends the most important part of his life after death, the period spent there lengthening as he evolves.

The three bodies in which he functions in these worlds are all mortal; they are born and they die. They improve life after life, becoming more and more worthy to serve as the instruments of the unfolding Spirit. They are copies in denser matter of the undying spiritual bodies, which are unaffected by birth and death, and from the clothing of the Spirit in the higher worlds, wherein he lives as the spiritual Man, while he lives here as the man of flesh, the "carnal" man. "These undying spiritual bodies are that of which S. Patil speaks: "We know that if our earthly house of this tabernacle were dissolved we have a building of God, an house not made with hands, eternal in the heavens. For in this we groan, earnestly desiring to be clothed upon with our house which is from heaven".[11]

The three mortal bodies are: the physical, the astral, and the mental, and they are related severally to the three worlds above-named.

The Physical Body

This is at present the most highly-developed body of man, and the one with which we are all familiar. It consists of solid, liquid, gaseous and etheric matter, the first three exquisitely organised into cells and tissues, these being built into organs which enable the consciousness to become aware of the outside world, and the latter possessing vortices through which forces pour. As the etheric part of the body separates at death from the solid, liquid, and gaseous part, the physical body is often subdivided into dense and etheric; the former is composed of the organs which receive and act; the latter is the medium of the life-forces and their transmitter to its dense comrade. Any tearing of the etheric part from the dense body during physical life is unwholesome; it is torn out by anaesthetics, and slips out, undriven, in some peculiar organisations, generally termed "mediumistic"; apart from its denser comrade, it is helpless and unconscious, a drifting cloud with force centres, useless when there is nothing to which it can transmit the forces playing through it, and subject to manipulation by outside entities, who can use it as a matrix for materialisation. It cannot go far from the dense part of the body, since the latter would perish if disconnected from it; when disconnection occurs the dense part "dies"—i.e., loses the inpouring of the vital forces which sustain its activities—even then the etheric part, or etheric double, hovers near its life-partner, and is the "wraith," or "shade," sometimes seen after death, drifting over graves. The physical body as a whole is man's medium for communication with the physical world, and it is sometimes called, for this reason, the "body of action." It receives also vibrations from the subtler worlds, and when it is able to reproduce these, it "feels" and "thinks," its nervous system being organised to reproduce these in physical matter. As the viewless air, strongly vibrating, throws the denser water into ripples, as the viewless light throws the rods and cones of the retina into activity, so does the viewless matter of the subtler worlds throw into responsive vibrations the denser matter of our physical body, both etheric and dense. As evolution proceeds, and the physical body evolves—i.e., appropriates finer and finer combinations of matter from the outside world—it becomes responsive to more and more rapid vibratory waves, and the man becomes more and more "sensitive". Racial evolution largely consists in this ever-increasing sensitiveness of the nervous system to outside impacts; for health, the sensitiveness must remain within limits of elasticity—i.e., the system must immediately regain its normal condition after distortion—if this condition be present, such sensitiveness is on the crest of the evolutionary wave, and makes possible the manifestation of genius; if it be not present, if equilibrium be not swiftly and spontaneously resorted, then the sensitiveness is unhealthy and mischievous, leading to degeneration finally, if unchecked to madness.[12]

The Astral Body

The development of this body differs enormously in different persons, but in all it is the body which yields the experience of pleasure and pain, which is thrown into action by passion, desire and emotion, and in which reside the centres of our sense-organs—of sight, hearing, taste, smell and touch. If the passion, desire and emotion are low, sensual animal, then its matter is coarse, its vibrations consequently are comparatively slow, and its colours are dark and unattractive—browns, dark reds and greens, and their combinations lit from time to time with flashes of scarlet.[13]

We are using this body throughout our waking hours, and in educated and refined people it has reached a fairly high stage of evolution. Its finer matter is closely in touch with the coarser matter of the mental body, and the two are constantly working together, acting and reacting on each other. In sleep the astral body slips out of the physical body. We learn much during our sleep, and knowledge thus gained slowly filters into the physical brain, and is occasionally impressed upon it as a vivid and illuminative dream. For the most part the consciousness in the astral world concerns itself little with the happenings there, being chiefly interested in its own exercise is thought and feeling; but it is possible to turn it outwards, and to gain knowledge of the astral World. Communication with friends who have lost their physical bodies by death is constantly carried on there, and the memory may be brought back into waking consciousness, thus bridging the gulf otherwise made by the death.[14]

Premonitions, presentiments, the sensing of unseen presences and many allied experiences are due to the activity of the astral body and its reaction on the physical; their ever-increasing frequency is merely the result of its evolution among educated people. In a few generations it will be so generally developed that it will become as familiar as the physical body. After death we live for some time in the astral world in the astral body used during our life on earth, and the more we learn to control and use it wisely now the better for us after death.

The Mental Body

This body, of finer material than the astral as the astral is fitter than the physical, is the body which answers by its vibrations to our changes of thought. Every change in thought makes a vibration in our mental body, and this, transmitted by the astral to the physical, causes activity in the nervous matter of our brains. This activity in the numerous cells causes many electrical and chemical changes in them, but it is the thought-activity which causes these, and not the changes which produce thought, as the materialists of the nineteenth century imagined.

The mental body, like the astral, varies much in different people; it is composed of coarser or of finer matter, according to the needs of the more or less unfolded consciousness connected with it. In the educated it is active and well-defined; in the undeveloped it is cloudy and inchoate. Its matter, drawn from the mental plane, is that of the heaven-world, and it is continuously active, for man thinks in his waking consciousness when out of the physical body in sleep, and after death, and lives wholly in thought and emotion when he leaves the astral world behind him, and passes into heaven. As this is the body in which long centuries will be passed in the heaven-world, it is only rational to try to improve it as much as possible here. The means are study, thought, the exercise of good emotions, aspiration (prayer) and beneficent endeavours, and, above all, regular and strenuous meditation. The using of these will mean a rapid evolution of the mental body, and an immense enrichment of the heavenly life. Evil thoughts of all kinds befoul and injure it, and, if persisted it will lead to veritable diseases and maimings of the mental body, incurable during its period of life.

Such are man's three mortal bodies: he casts off the physical at death, the astral when ready to enter the heaven-world. When he has finished his heaven-life his mental body also disintegrates, and he is a Spirit clad in his immortal bodies. On descending for re-birth a new mental body is formed and a new astral, conformable to his character, and these, attach themselves to his physical body, and he enters by birth in a new period of mortal life.[15]

Man's Immortal Bodies

"We have a building of God, an house not made with hands eternal in the heavens," said the great Christian Initiate S. Paul, " for in this (body) we groan, earnestly desiring to be clothed upon with our house which is from heaven." This heavenly house it is which is built of Man's immortal bodies, the habitation of the Spirit through unending ages, the dwelling-place of man himself, through births and deaths, through the measureless period of his immortal life in manifestation.[16]

The Spirit which is "the offspring of God" abides ever in the bosom of the Father, in very truth a Son of God, sharer in His eternal life. God made man "to be the image of His own eternity." This Spirit we call the Monad, because it is a unit, the very essence of Self-hood. The Monad, when he descends into matter in order to conquer and spiritualise it, appropriates to himself an atom of each of the three higher worlds, to make the nucleus of his three higher bodies—the super-spiritual, the spiritual and the intellectual. To these, by a thread of spiritual, (*buddhic*) matter, he attaches also a particle from each of the three lower worlds, the nucleus of the three lower bodies.

For long, long ages he broods over these, as his future mortal bodies, just touched with his life, climbs slowly upwards through the mineral, vegetable and animal kingdoms; meanwhile little aggregations of the matter of the three higher worlds, the "building of God ... in the heavens" form a channel for his life, beginning to manifest in those worlds; and when the animal form has reached the point at which the upclimbing life makes strong appeal to the higher, he sends down through these an answering pulse of his life, and the intellectual body is suddenly completed, as light flashes out between the carbons of an electric arc. The man has individualised for life in the lower worlds.

The super-spiritual (atmic) body is but an atom of its lofty world, finest film of matter, embodiment of Spirit, "God made flesh" in a very real sense, divinity dipping down into the ocean of matter, not less divine because embodied. Gradually into this super-spiritual body will pass the pure result of all experiences, stored up in eternity, the two lower immortal bodies gradually merging themselves in it, blending with it, the glorious vesture of a man consciously divine, made perfect.

The spiritual (*buddhic*) body is of the second manifested world, the world of pure spiritual wisdom, knowledge and love in one, sometimes called the "Christ-body," as it is this which is born into activity at the first great Initiation, and which develops to the "fullness of the measure of the stature of Christ" on the Path of Holiness. It is fed by all lofty and loving aspirations, by pure compassion and all-embracing tenderness and pity.[17]

The intellectual (causal) body is the higher mind, by which the man deals with abstractions, which is "of the nature of knowledge," in which he knows truth by intuition, not reasoning, borrowing from the lower mind ratiocinative methods only to establish in the lower world abstract truths which he himself knows directly. The man in this body is called the ego, and when this body blends with the next above it, he is called the spiritual ego, and begins to realise his own divinity. It is fed and developed by abstract thinking, by strenuous meditation, by dispassion, by the yoking of intellect to service. It is by nature separative, being the instrument of individualisation, and must grow strong and self-sustaining in order to give the necessary stability to the subtle spiritual body with which it is to blend.

These are man's immortal bodies, subject neither to birth nor death; they give the continuous memory which is the essence of individuality; they are the treasure house of all that deserves immortality; into them can enter "nothing that defileth"; they are the everlasting dwelling-place of the Spirit. In these the promise is realised: "I will dwell in them and walk in them." These fulfil the prayer of the Christ: "That they also may be one in Us." These make good the triumphant cry of the Hindu: "I am Thou".[18]

Steps on the Path

The normal course of human evolution leads man upwards, stage by stage. But an immense distance separates even the genius and the saint from the man "stands on the threshold of divinity"—still more from him who has fulfilled the Christ's command: "Be ye perfect, as your Father in heaven is perfect." Are there any steps which lead up to the gateway of which it is written: "Strait is the gate and narrow is the way that leadeth unto life, and few there be that find it?" Who are "this perfect," of whom Paul the Apostle speaks?

Truly are there steps which lead up to that Portal, and there are few which tread that narrow way. The Gate is the Gate of Initiation, the second birth, the baptism of the wholly Ghost and of Fire; the Way leads to the knowledge of God, which is life in the Eternal.

In the western world the stages, or steps have been called: Purgation, Illumination, Union; by those stages the Mystic—who is rapt to the Beatific Vision by devotion—denotes the Path. In the eastern world the Occultist—the Knower, or Gnostic—sees the steps in somewhat other fashion, and divides the path into two great stages, the Probationary and the Path Proper; the Probationary represents the Purgation of the Mystic: the Path itself the Mystic's Illumination and Union. He further seeks to develop in himself on the Probationary Path certain definite "qualifications," fitting him to pass through the Portal which ends it; while on the Path itself he must wholly cast away ten "fetters," which hold him back from attaining Liberation, or Final Salvation, and must pass through four other Portals, or Initiations.[19]

The qualifications must each be developed to some extent, though not completely, ere the first Portal can be passed. They are:

(1) *Discrimination:* the power to distinguish between the real and the unreal, the eternal and the fleeting—the piercing vision which sees the True and recognises the False under all disguises.

(2) *Dispassion or Desirelessness:* the rising above the wish to possess objects which give pleasure or to drive away objects which give pain by utter mastery of the lower nature, and transcending of the personality.

(3) *The Six Endowments, or Good Conduct:* control of the mind, control of the body—speech and action—tolerance, endurance or cheerfulness, balance or one-pointedness, confidence.

(4) *Desire for Union, or Love:* These are the qualifications, the development of which is the preparation for the first Portal of Initiation. To these should the man address himself with resolution, who has made up his mind to travel forward swiftly, so that he may become a Helper of Humanity.

When he has acquired sufficient of these so to knock the Door that it shall be opened unto him, he is ready to pass over its threshold and to tread the Path. He is initiated, or receives the "second birth." He is called among the Hindus the Wanderer (*Parivrajaka*), among the Buddhists "he who has entered the stream" *Srotapanna* or *Sotapanna*: and before he can reach the second Initiation he must cast off wholly the "fetters" of Separateness—he must realise that all selves are one; Doubt—he must know and not merely believe the great truths of *Karma*, Reincarnation and the Perfection to be reached by the treading of the Path; Superstition—the dependence on rites and ceremonies. These three fetters wholly cast off, the Initiate is ready for the second Portal, and becomes the Builder (*Kutichaka*), or "he who returns but once," Sakadagamin; he must now develop the powers of the subtle bodies, that he may be useful in the three worlds, fitted for service. The passing through the third Portal makes him the United (Hamsa, "I am He"), or "he who does not return" save with his own consent—the Anagamin. For the fourth Gate should be passed in the same life, and for him who has passed that, compulsory rebirth is over. Now he must throw off the fetters of Desire—such rarefied desire as may be left in him and of Repulsion nothing must repel him for in all he must see the Unity. This done, he passes through the fourth Portal, and becomes the Super-individual (Paramahamsa, "beyond the I") or the Venerable (Arhat). Five are the filmy fetters that yet hold him, and yet so hard is it to break their cobweb subtlety that seven lives are often used in treading the space that separates the Arhat from the Master, the Free, the Immortal, the Super-Man, "He who has no more to learn", in this system, but may know what He will by turning on it His attention. The fetters are Desire for life in form, desire for life in formless worlds, pride—in the greatness of the task achieved, possibility of being disturbed by aught that may happen, illusion—the last film which can distort the Reality. When all these are cast away forever, then the triumphant Son of Man has finished His human course, and He has become "a Pillar in the Temple of my God who shall go out no more"; He is the Man made perfect, one of the First-born, an Elder Brother of our race.

We have traced the steps by which a man may climb to the status of the super-man. Let us now consider the relation to the world of Those who stand at that great height, and who yet are of the human family, our Elder Brothers.

All religions look back to a founder, who rose high above humanity; all ancient history tells, of lofty Beings, who laid the foundations of nations, and guided them during their infancy and youth. We hear of divine Kings, of divine Dynasties, of divine Teachers; the testimony of the past is so unanimous, and the ruins remaining of past civilisations are so mighty, that we cannot reasonably declare the testimony to be worthless, nor the civilisations to be the unaided product of an infant humanity.

It is also noteworthy that the most ancient Scriptures are the noblest and most inspiring. The Classic of Purity of China, the Upanishads of India, the Gathas—fragmentary as they are—of Persia, are far above the level of the later religious writings of the same countries; the ethics found in such ancient books are all authoritative, not hortative, they teach "as having authority, and not as the scribes."

No religion denies or ignores these facts, as regards its own Teachers and its own Scriptures; but unhappily, most are apt to deny or ignore them where the Teachers and the Scriptures of other religions are concerned. Students of the Wisdom realise that all these claims must be impartially recognised or impartially rejected; and Occultists know that while many legends and fables may have gathered round these mighty Beings, none the less They, of a verity, have existed in the past and exist in the present.

The Occult Hierarchy which rules, teaches and guides the worlds is a graded Order, each rank, having its own multifarious duties and carrying them out in perfect harmony, working out a portion of the plan of the Supreme Lord, the Logos of the system in a service which is "perfect freedom". Two leading departments of our section of this Hierarchy are concerned, the one with the ruling, the other with the teaching of our worlds.

Those whom the Hindus call the four Kumaras are the Chiefs of the Ruling Department, and the Manus of Rounds and Races are Their Lieutenants, with, below Them, the grade of Adepts, which numbers among its members Those called Masters, to carry out the details of Their work. Theirs to guide evolution, to shape races, to guide them to continents builded for their dwelling to administer the laws which cause the rise and fall of peoples, of empires, of Civilisations.

At the head of the Teaching Department stands the "Enlightened," the Buddha, who, when He passes away from earth, hands the Teacher's staff to Him who is to become a Buddha in His turn, the Bodhisattva, the actual Teacher of the worlds. This Supreme Teacher is the ever-living Presence who over-shadows and inspires the world-faiths, who founds them, as they are needed for human guidance, and who, through His Helpers among the ranks of Adepts, guides each religion so far as is permitted by the stubbornness and ignorance of men. Every great spiritual wave flows from this department of the White Brotherhood, and irrigates our earth with the water of life.

In the grade of Adepts alluded to above are Those to whom the name of "Masters" more peculiarly belongs, in that They accept as *chelas*, or disciples, those who have reached a point of evolution fitting them to approach the Portal of Initiation, and are resolutely striving to develop in themselves the qualifications before described. There are many of this rank in the Hierarchy—Those who have passed the fifth Initiation—who do not take pupils, but are engaged in other work for the helping of the world. Even beyond this rank some will still keep under Their charge chelas who have long been devoted to Them, the tie formed being too sacred and too strong to break.[20]

REFERENCES

1. Besant, Annie, *The Riddle of Life* (Adyar: The Theosophical Publishing House), 1955.
2. *Ibid.*, p. 2.
3. *Ibid.*, p. 4.
4. Besant, Annie, *The Ancient Wisdom. An Outcome of the Theosophical Feelings* (Adyar: The Theosophical Publishing House), 1955, pp. 66-70.
5. Nethercot, Arthur H., *The Early Four Lives of Annie Besant* (London: Rupert Hart Davis), 1961, p. 303
6. Besant, Annie, *The Riddle of Life, op. cit.*, pp. 8-9.
7. Besant, Annie, *The Ancient Wisdom. An Outcome of the Theosophical Feelings, op. cit.*, p. 278.
8. *Ibid.*
9. Besant, Annie, *The Riddle of Life, op. cit.*, pp. 11-12.
10. Nethercot, Arthur H., *The Early Four Lives of Annie Besant, op. cit.*, p. 305
11. Besant, Annie, *The Riddle of Life, op. cit.*, p. 5.
12. Besant, Annie, *The Ancient Wisdom. An Outcome of the Theosophical Feelings, op. cit.*, pp. 43-62.
13. Leadbeater, C.W., *Man Visible and Invisible* (Adyar: The Theosophical Publishing House), n.d., p. 65.
14. *Ibid.*, pp. 63-80.
15. *Ibid.*, pp. 118-150.
16. *Ibid.*
17. *Ibid.*
18. Besant, Annie, *The Riddle of Life, op. cit.*, pp. 25-26.
19. *Ibid.*, p. 62.
20. *Ibid.*, pp. 62-68.

4

HOME RULE MOVEMENT

The opening years of the twentieth century, when the Indian National Congress was hardly one-and-a-half decade old, witnessed some controversial administrative measures undertaken by the Viceroy Lord Curzon. These measures, otherwise thought to be the measures for the smooth functioning of the British rule in India, raised serious controversy and agitational approach in Bengal and elsewhere, and caused an anti-Raj stance which flared up in the minds of the extremists and moderates of the time. This kind of agitation was not seen earlier, at least after the events of 1857. This factor created much awakening amongst the people of India and made them quite conscious of their constitutional rights.

During this period two groups of politicians with different ideologies and methods for achieving their goal emerged at an all-India level. On account of their moderate views and the liberal constitutional style, one group was termed as the moderates, and the other as the extremists due to their approach mingled with militant and extremist style. The ideology of these two political groups greatly shaped the programme of the Indian National Congress in the coming few years which witnessed numerous changes at the administrative levels, both in India and Britain. The ideology, whether of moderates or extremists, did effect the working of the mind of British bureaucracy in India which through its regular flow of official cablegrams to the top functionaries of the Home government made them realize the inadequacy of the constitutional concessions granted to Indians in the past, thereby recommending something more to be granted through the agency of the House of Commons.

Inspired and influenced by the Gladstonian political style of liberalism which dominated the political scene in Britain for the last three decades of the nineteenth century, the early Congress leaders were quite moderate in their demands as they had realised the efficiency of the Raj in one way or the other, and they formulated a strong feeling to the fact that the British rule alone could secure to the country the peace and order which were necessary for slowly evolving a nation out of the heterogeneous elements of which it was composed, and for ensuring to it a steady advance in different directions.[1]

Undoubtedly the moderates considered our connection with the British as providential. They had the conviction that with British help, sympathy and support, India would recover her national freedom. In the thirteenth session of the Indian National Congress, it was clearly stated, 'It should not be forgotten for a moment that the real link that binds us to England is the hope, the well-founded hope and belief, that with England's help we shall, and under her guidance alone, we can attain national unity and national freedom.'[2]

The insistence on gradualness for the attainment of the desired goal was plank of thought on which the moderates harped their working of mind. The feelings of this kind were expressed by Gopal Krishna Gokhale in his presidential address in the Congress session held at Banaras in December 1905.[3] He in fact emphasised with full support of his like-minded colleagues, the need of gradualness for the attainment of self-government within the British Empire.[4] He said that 'the proper exercise of the political institutions of the West can be acquired by an Eastern people through practical training and experience only....' He believed that Indians must be renovated, purified and perfected before India could be the master of her own destiny.[5]

Significantly the composition of the class of moderates had deep linkage mainly with the elite class. They had the tradition of learning since generations during the Raj as well as prior to it. Some of them were big landlords and zamindars owning large estates in the rural areas, but these families were mostly settled in the urban areas. The reason was obvious. They had numerous facilities to organize and participate in the political, social, cultural, religious and educational activities of the town-life. Some of them who could have the privilege of higher education began to show prominence in the field of law, medicine, education and journalism. Indeed a few of them were instrumental in having a deep imprint on the minds of the people through their writings which had Indian nationalism as their major theme. This kind of traditional training had made them to adopt a new model of putting forth their demands before the British bureaucracy, and it was mainly through prayer, petition and mild type of protest.

Thus the ideology of the urban-based moderates with the basic facilities of the town-life was instrumental in moulding the political and social life of the people in the early years of the twentieth century. No doubt they adhered to the Indian traditions from which they derived much inspiration. They were also influenced by the political philosophy of John Stuart Mill, Thomas Paine, Adam Smith, Herbert Spencer, Muzzini, John Bright, Edmund Burke, and last but not the least, William Edward Gladstone.

Almost in every session of the Indian National Congress, the demand for constitutional concessions for India was repeated, i.e. its resolutions which had the unanimous support of all the delegates assembled under its flag. The major demand was worded as colonial self-government within the British Empire. To achieve this goal, they recommended constitutional methods. They banked upon 'the advice of English liberals' rather than bringing pressure from below to expedite the pace of political progress in India.

The Indian National Congress was controlled and managed by the prominent moderate leaders between 1908-1915, and the annual sessions were presided over by men like Madan Mohan Malaviya, William Wedderburn, Bishan Narayan Dhar, Rashbehari Ghose, R.N. Madholkar, Bhupendra Nath Basu, Nawab Syed Mohammad and Satyendra Pritsud Sinha. They demanded more and more representation for Indians in the Indian Councils. The other subjects discussed and put in as demands before the government of India were better deal in excise administration, respectable treatment to Indians in British colonies, periodical inquiry into the conditions in India to enable the parliament to discharge more satisfactorily its responsibility in regard to the government of India and the change in the existing economic relations between India and Britain.

The attitude of the moderates towards the extremists was that of alarm. They thought that the extremists were challenging the very basis of the older thought and belief. During this period, they 'disowned' the extremists as serious differences cropped up between the two groups.[6] The moderates opined that the extremists were bent upon endangering national progress by their militant actions and policy.

On the other hand, the extremist group in the Indian National Congress during this period was led by prominent leaders like Lajpat Rai, Bal Gangadhar Tilak, Bepin Chandra Pal and Aurobindo Ghose. Their line of thinking and the practical politics were greatly influenced by the teachings of Swami Dayananda and Vivekananda. In fact, when Lajpat Rai came back to India from England, he felt much disillusioned after making a thorough assessment of the working of British mind towards the grievances of Indians.[7]

The programme of the extremists was an anti-thesis of the moderates programme. They dismissed the idea of prayer and appeal in any form to the government as they labelled it as useless mendicancy. They criticised the methods of moderate leaders and opined that it would be the force of circumstances which alone would compel the British to hand over power to Indians.

The Surat split in the Congress in 1907 led to serious allegations of one group against the other. As a result thereof, the extremists of one group were against the other. The younger and extremist section in the Indian National Congress broke away from the party. Lajpat Rai was very critical of the moderates and called them hypocrites who dragged Indian patriotism into the mire. So far so, Bipin Chandra Pal denied the feasibility of a self-governing India within the Empire on the ground that India constituted the pillar of the Empire, and therefore, could exert for effective self-government only if the Empire itself ceased to exist. Aurobindo Ghose opined that demanding concessions from the government was futile and irrelevant, and wished to have a free national government unhampered in the least degree by foreign control.

But after the split at Surat, the extremists failed to build up any effective organization of their own which might have enabled them to maintain their identity on a firm footing. The imprisonment of Tilak from 1908-14; the retirement to Pondicherry by Aurobindo Ghose and numerous coercive measures undertaken by the government did much damage to the extremists.

During this period, a new kind of feeling had developed in the minds of the people of India which had glaringly apprised them of numerous weaknesses in the international spheres which had greatly effected the Raj in its colonies. The international situation had taken such a shape that it became difficult for her to boast of even a few friendly nations in the European hemisphere. Obviously, in France and Russia, she had her traditional rivals, and Germany under the new Kaiser had also entered into perpetual rivalry. Besides, the military and naval preparations of Germany were phenomenal. Britain's diplomatic isolation had become a glaring phenomenon during the Boer War which touched upon a few vital regions of Europe. Above all, the defeat of Russia by Japan, which was thought to be something like unexpected phenomenon had greatly enthused some Asian powers which gained more and more hope, confidence and renewed strength for a better future of their country. The urge for gaining independence from the foreign yoke gained momentum and India was indeed no exception. At this point of time, India developed the feeling that if a small nation like that of the Japanese could show so much unity, well-knit organised action and strength against a huge and militarily well-developed country like Russia, it could also be possible for her to achieve this kind of success over the Raj by devising and implementing means, revolutionary in contents.

To placate the new upsurge of Indians a few constitutional concessions were announced in 1909. The Minto-Morley Reforms had an historic setting. Numerous discussions took place at various levels; views and statements of prominent leaders were recorded and assessed, and varied political ideas enumerated in different political style were accommodated. The curtain, however, was rung down on the parliamentary drama when on 25 May 1909, the debates in the House of Commons took the final shape and it set its seal of approval on Morley's India Council Bill and placed it on the Statute Book. In fact the councils were enlarged and their functions extended. They comprised in part of members elected by the media of constituencies—small and narrow indeed. In the provincial councils, the position was changed slightly in favour of Indian members as their number was increased.

'In fact, one of the main purposes of reforms was to create counterpoise to the Congress which stood and worked for national unity. Both Morley and Minto were keen on rallying the loyal elements...that is, the groups which wanted to perpetuate British rule in India, support the foreign government in maintaining its hold upon the country, and to discourage and indeed obstruct the educated class, which was engaged in the tremendously difficult task of building a nation....'[8]

Minto explained earlier that '...constitutional autocracy binds itself to govern by rule, admits and invokes to its council representatives of all the interests which are capable of being represented. (But) it reserves to itself the predominant and absolute power.'[9] In fact the aim of reforms in accordance with the views expressed both by Morley and Minto was greater association of Indians with the government. In a brief communication with Minto, Morley expressed his anxiety thus on 16 May 1907, 'We are all in a cleft stick, we don't know the minds of the Natives, and the Natives don't know what is in our minds. How to find some sort of a bridge? That's the question.[10]

Thus Minto reacted and came to the conclusion that the principle of autocracy and the principle of constitutionalism might be fused in a well-planned and well-organised system of constitutional autocracy. The success of autocracy, in order to govern and administer well, was to be maintained by crushing the aspirations of the leaders of the party which, time and again, put forward the plea for self-government and at the same time by adopting the policy of conciliation of the aristocratic elements and moderate men who 'will range themselves on the side of the Government and will oppose any further shifting of the balance of power and any attempt to democratise Indian institutions.'[11]

The obvious reaction of the Conservatives in the House of Commons was the same as was expected. They regarded it as a revolutionary measure, wholly unnecessary and harmful. Arthur Balfour, the leader of the opposition representing the Conservative Party referred to Morley's remarks about the unfitness of Indians for representative government in a critical way, 'Why you should do everything to make assemblies which are not representative, and which you do not intend to make representative, makes of all the worst and most laborious parts of our procedure, that I admit absolutely passes my comprehension.'

In December 1909, when the Indian National Congress held its session at Lahore under the presidentship of Pandit Madan Mohan Malaviya, he expressed his shock at Morley's surrender on the question of Muslim representation. He opined, 'This is protecting the interests of a minority with vengeance. It looks more like a case of allowing the majority to be driven to a corner by a minority ... The Hindu minorities in the two provinces (Punjab and Eastern Bengal and Assam) have been left out severely in the cold.... The regulations for the first time in the history of British rule have recognised religion as a basis of representation, and have thus raised a wall of separation between the Mahommedan and non-Mahommedan subjects of His majesty.[12]

At this point of time, the important nationalist leaders were either behind the bars or in exile. The chain of violence and repression did not subside; rather doubts and apprehensions continued to mount towards the direction which, one day, was to take the shape, colossal in magnitude. As Valentine Chirol informed Lady Minto on 4 May 1910, 'The Reform, likewise are a palliative which may for a time keep the political opportunists on the fence. But they do not touch the root of the evil.'

The reforms, however, did not concede real constitutional rights. Their emphasis upon the principles of community and group interests as the basis for the future political arrangements in India came in for severe criticism and condemnation by one and all except those who were the beneficiaries on account of the British sympathy, help and patronage. Indeed the principle of separate representation conceded in the reforms was tantamount to the denial of Indian nationalism. The Morley-Minto Reforms thus created a gulf which was never bridged.

The new wave depicting fervent zeal of patriotism had already been enkindled in the hearts of the young men of India. Initiative began to pass from Britain to India. Minto admitted that 'we all feel that we are mere sojourners in the land, that we are only camping and on the march. Everyone is looking forward to a well-earned rest at home at last'.[13] Morley also showed a sense of pessimism so far as the strength and lasting capability of the Raj was concerned. He opined, 'No world had lasted for ever and was doubtful whether the British empire had many years of life before it...the longer one stares at India, the more grim and ugly it all is.'[14] It was probably so, as he realized through a regular communication with the Viceroy that there was growing estrangement between the British bureaucracy and the Indian leaders who got themselves seriously engaged in numerous political commitment with the masses of India.

The activities of a newly emerged Ghadr party in the Punjab with its deep linkages in some countries abroad had inculcated much awakening amongst the peasantry and the lower middle class. With the passage of time, the Ghadrites had begun questioning the role of the Raj, mainly the performance of its bureaucracy in the existing set up. This kind of resentment and opposition was in terms of the benefits enjoyed by the people of India as a whole under the British government. Initially, the initiative in this regard was taken by Lala Har Dayal and Bhai Parmanand, who were responsible for imparting political education to Indians settled in Vancouver and San Francisco. Their well-known paper *Ghadr,* published in Punjabi and Urdu, was financed through a generous collection by a sizable section of Indian society. This paper reached the hands of numerous Punjabi and Urdu knowing population which greatly influenced the minds of the readers. Its articles and editorials had anti-Raj stance. Today there begins in foreign lands, but in our country's language, war against the English Raj. What is our name? Mutiny! *Ghadr* means mutiny. What is our work? Mutiny!, Where will mutiny break out? In India... The time is coming soon when rifles and blood will take the place of pen and ink. Bravemen and worthy sons of India! Be ready with bullets and shots. Soon the fate of the tyrant will be decided and days of happiness and glory will dawn in India!'

The *Ghadr* party had thus a revolutionary programme which had the following objects in view, viz., to seduce Indian troops, to murder loyal subjects and officials, to loot banks and government treasury, to make propaganda amongst the masses through distribution of seditious literature; to commit dacoities; to procure arms and ammunition; to manufacture bombs, hand-grenades and other weapons; to establish secret societies for propaganda of the *Ghadr* party's programme, to loot police establishments, to destroy railways, telegraph offices and other government establishments, and last but not the least, to mobilise the people living in the villages.

Indeed the tragic fate of the passengers of the *Komagata Maru* who landed on the sea-shores of India, was an historic event. Some of them were shot there and then, and others were arrested and convicted to various terms of imprisonments. Those who escaped this tragic drama, engaged themselves in the revolutionary activities during the First World War. They made a serious effort to provoke mutiny among the Indian troops. Seditious pamphlets were systematically distributed in many regions of the Punjab. Secret emissaries were sent to the Meerut Cantonment in order to kill European officials there.

These revolutionary activities of the *Ghadrites* had nightmarish effects on bureaucracy. As a measure of precaution, severe action was taken against them. Their haunts and rendezvous were consistently raided by the police parties, and members of the party were arrested and convicted. The punishments awarded were from deportation to hanging. On account of these repressive measures, the movement could not make much headway. But it did leave a significant impact on the people of the Punjab regarding the efficacy of the Raj, and a feeling tinged with an *anti-Raj* sentiments was imbibed amongst the uneducated and patriotic peasantry and lower middle class which could be mobilized at a time, when the province as a whole was to fight for a common cause for whole of India at later stage.

The outbreak of the First World War in August 1914 worked as a 'catalyst' to the growing political aspirations of Indians. The diplomacy of Germany in Europe and elsewhere had far reaching effects at global level in the first two decades of the twentieth century. The war was the result of economic rivalries and competition for colonial expansion between the European powers. Germany's rapid industrial development and enormous growth of army, navy and numerous other defence devices posed a serious challenge to the political and economic supremacy of England, not only in India but in other regions where it had deep-rooted interests. In such circumstances, preparations of highest level on the part of the British could be a match against German preparations.

In this global involvement mingled with numerous uncertainties on land and sea, the cooperation, help and sympathy of the British colonies in various regions of the world were of prime significance. It naturally wanted India to contribute as much as it could for the cause of the world war. This could only be possible if the British government in India was relieved of *anti-Raj* activities of the people of our country. And at the same time devoted efforts were made at various levels to help the government in men and material.

The appeals at various ends—from loyalists to those who realised the difficulties of the Raj—were much effective in content. The same year, when the annual session of the Indian National Congress was held in Madras in the month of December, it was resolved, 'to convey to His Majesty the King-Emperor and the people of England its profound devotion to the throne, its unswerving allegiance to the British connection, and its firm resolve to stand by the empire at all hazards and at all costs.'[15]

India's participation in the First World War did not provoke any kind of systematic resentment amongst the people. The spirit of help, sympathy and cooperation was glaringly visible in all the regions. The nationalists, however, hoped that the political situation after the war might turn to their advantage culminating into some constitutional concessions. With this hope in view, the leaders of the Congress party made no secret of the fact that they would have some concessions from the Raj as a price of their genuine help of men and money to it. In the annual sessions of the Indian National Congress held in 1914-15 they declared that 'in view of the profound and avowed loyalty that the people of India have manifested in the present crisis, the time has arrived to introduce further and substantial measures of reform toward the attainment of Self-Government.'[16] The only notable exception was Mahatma Gandhi who whole-heartedly participated in the war efforts, and became almost a 'recruiting agent of the Government without any condition...'[17]

So far so, even Bal Gangadhar Tilak who was regarded by the government 'as the inveterate enemy of British rule', declared at such a crisis: 'It is the duty of every Indian, he be great or small, rich or poor, to support and assist His Majesty's Government to the best of his ability.'[18]

It is quite evident that the Muslims were unhappy at the attitude of the western powers, especially Britain, towards Turkey. But they too demonstrated a sense of loyalty to the crown during the war of zealously fighting against the Turks in Gallipoli and Mesopotamia.[19] By doing so, they 'demonstrated how little hold the religious leaders, Maulvis and Mullahs, had on their mind.'[20]

The other sections of Indian society—the state, rulers, landlords, zamindars, business magnates and non-political classes vied with the politicians in joining the chorus of loyalty.[21] It is true that in different communities, interests and classes, the motivation was different, differing from genuine traditional reverence for the ruling authority or sarkar, to expectation of communal reward or national gain. This pro-British stance was in many ways helpful in the prosecution of the war. It gave the government confidence in denuding India of both British and Indian troops and sending them to the front. It aided the effort to raise from India a large contribution in men, money and material. It guaranteed the failure of German plans in support of the revolutionary movement.'[22]

The voluntary response afforded by various sections of Indian society during the war was unprecedented in the annals of India. It was indeed unexpected in the circles of British bureaucracy both in England and India. They were surprised at this and the Prime Minister Asquith acknowledged in the House of Commons that, 'in all the moving exhibitions of national and imperial patriotism which the war had evoked, there was none which had more touched the feelings of the British people than the magnificent response which the princes and the people of India had made to their need'.[23]

Indian help for war was also appreciated by Charles Roberts, Under Secretary of State for India on 26 November 1914, who described it as '...this striking and historic event—the participation of India in force in the World War of the Empire'.[24] He opined, '...but it is clear that India claims to be not a mere dependent but a partner in the Empire, and her partnership with us in spirit and on battlefield cannot but alter the angle from which we shall all henceforward look at the problems of the government of India'.[25] He hoped that 'the common endeavours of these days would enable India to realise that she was occupying, and was destined to occupy a place in our free Empire worthy of her fighting races and the patriotism of her sons.'[26]

This kind of active participation in the war at a large scale engendered a sense of pride in the minds of Indians who began to feel that an Indian sepoy could fight with appreciable capability and shoulder to shoulder with a British soldier. The myth of the superiority of a white soldier was thus exploded. 'Ideas of equality of races and partnership in a free empire began to be entertained as practical issues.' As a result of these developments, moderates began to put forward pleas for declaring self-government as the goal of India's progress. Leaders like Mrs. Annie Besant and Bal Gangadhar Tilak began to raise the demand for Home Rule.

At this point of time, Indian politics was mainly organised and dominated by such politicians who were mostly western educated. The political scene was controlled by the moderate politicians who adhered to constitutional methods. The British government, therefore, favoured them. On the other hand, the extremists who believed in militancy were under the severe government repression. Political activities were confined to a few regions. But Annie Besant, who was influenced by Gladstonian liberalism, writings of John Stuart Mill, and, above all, the basic tenets of democracy, was neither a moderate nor an extremist. She was an individualist in politics and expressed dissatisfaction with the routine-like political activities of the Indian National Congress. She wished to have a quick result-oriented political activities which induced her to launch the Home Rule Movement. Besides, she launched a few subordinate organisations like 'Sons of India', 'Brothers of Service' and

'Youngmen Indian Association,' in order to implement her political, social and cultural programme in an effective way.

The launching of an All India Home Rule League in September 1915 was a most significant phenomenon in her career. Promptly she was instrumental in setting up a network of such organizations in various regions under the leadership of devoted and enthusiastic workers who were enjoined upon to push up the work of the Home Rule Movement amongst the masses. Soon numerous sections of people including peasants and workers became active participants under Besant's flag. 'India could never have been like it before—so one could say after witnessing such ferment and such passionate demonstration of active political life.'[27] Thus the year 1916 commenced with numerous significant developments which had deep linkage with the existing politics of the time. The country indeed looked poorer with the untimely demise of Gopal Krishna Gokhale and Pherozeshah Mehta. Bal Gangadhar Tilak had as yet no place in the Congress, and he had to wait a year more in order to find his place in the party. He made himself deeply involved in the Home Rule activities. The same kind of efforts were made by no less a leader than Annie Besant.

The objects of the Home Rule League, its definition and aims were explicitly explained thus.[28] The Home Rule has been established to carry out the following objects: To secure Home Rule for India through all law-abiding and constitutional activity; to maintain the connexion with Great Britain by becoming a free nation within the British Empire, under the Imperial Crown of His Majesty the King Emperor, George V and his successors; to support and strengthen the National Congress which has laboured for thirty years to lay the foundations of Indian self-government and to carry on a continuous educative propaganda on the necessity of Home Rule for India. In fact Home Rule means the government of the people and by the people.

'Home Rule means the right of the people to send their representatives to make laws for the people, and to levy takes on the people, subjection to laws not made by the representatives of the people is tyranny. The levying of taxes not approved of by the representatives of the people is robbery. Thus has England proclaimed, and she has enforced these principles in her own government, so that no one now denies them. Home Rule in England has carried out these principles. Home Rule in India will also carry them out.

'Home Rule' means the education of the people, free and compulsory for every child.

'Home Rule' means the cooperation of all to ensure a happy and prosperous life for every citizen of the state; the creation of conditions which will enable every child born into the nation to develop all the faculties, he or she brings with him or her into the world, so that each may grow to the full stature of the manhood or womanhood of which he or she is capable.

'Home Rule' means that every grown up person shall have a share in choosing the person who shall represent him in the village panchayat, the Taluq Board, the District Board, the Provincial Legislative Council and the Supreme Legislative Council. This will not all come at once. Every village resident householder will at first have a vote in the village panchayat; probably there will be a small property or educational qualification for the franchise for the higher Boards and Council. There will be no advantages or disadvantages attached to caste, creed or colour.

'Home Rule' means that white and coloured people will be equal before the law and in practice, that judicial officers, and their courts will be stationary, not peripatatic during the trial of a case; that the collector of revenue will not be a judicial officer also; that the judicial officer will not also be an executive officer.

'Home Rule' means that Indians will manage their own affairs, but foreign and international affairs, like peace and war, will be managed by a Council of the Commonwealth, to which India will send some of her own men like the other free nations belonging to it.

'Home Rule' means that the C.I.D. will no longer be allowed to meddle with political matters, to harass harmless people; to terrorize the law-abiding. It will be confined to its legitimate duty of hunting criminals, not patriots.

'Home Rule' means that the police will be the servants of the people, not their masters, will be feared by the bad citizen not by the good, and will have a pay sufficient to live upon without taking from the poor.

'Home Rule' means home defence. It means that, being free to manage your own land, you will also have to defend and protect your own land for safety. You will not be able to depend upon a foreign army, but you must maintain your own citizen army, and you must keep order within the motherland, as well as guard her borders.

'You will be a free nation in a free commonwealth. You will be men, not serfs; citizens, not dependents.'

In her charter of India's liberties, Annie Besant explained thus,

'To be free in India, as the Englishman is free in England;

'To be governed by her own men, freely elected by herself;

'To make and break ministries at her will;

'To carry arms, to have her own army, her own navy, her own volunteers;

'To levy her own taxes, to make her own budgets;

'To educate her own people;

'To irrigate her own lands, to mine her own ores, to mint her own coins;

'To be a sovereign nation within her own borders owing the paramount power of the Imperial Crown, and sendingher sons to the Imperial Council;

'Britain and India hand in hand; but an India free as is her right's.'[29]

In the Home Rule Section of the *New India*[30] issued on 15 June 1916.

A.O. Hume wrote a patriotic poem, which had much impact in India:

Sons of Ind, why ye sit idle
Wait ye for some Deva's aid?
Buckle to, be up and doing!
Nations by themselves are made.
What avail your wealth, your learning
Empty titles, sordid trade?
True Self-Rule were worth them all!

Nations by themselves are made.
Sons of Ind, be up and doing
Let your course by none be stayed
Lo ! the Dawn is in the East;
By themselves are Nations made!

'The cry of Home Rule spread to the remotest corners of the land and Home Rule Leagues were established all over the country. In Mrs. Besant's hands the press became powerful in spite of repression under the Press Law ... Another factor that largely contributed to the great success of the movement was that from its inception it recognised the integrity of language areas, and in organising the country, adopted the linguistic principle as determining the provincial delimitations.'[31]

The first meeting of the Home Rule League of Annie Besant was held in Madras on 3 September 1915.[32] As many as five hundred persons enrolled as members of the League. The All India Home Rule League initially had an Executive Council of seven office bearers who were elected for a brief period of three years by the thirty-four founding branches.[33] Obviously Annie Besant was elected its president and G. S. Arundale as its organizing secretary. Besides C.P. Ramaswami Aiyar was one of the general secretaries and B.P. Wadia the treasurer.[34] The functions of the League were conducted by these four officials from its headquarters at Adyar.[35]

The Home Rule League consisted of two divisions—one in India and the other in England. Whereas the function of the former was to educate Indians about the implications of the Home Rule Movement, the latter was to educate the English public opinion in relation to the political condition of India in the first two decades of the twentieth century.

Some of the moderates, however, did not like the idea of the birth of a political organization of this type as they apprehended 'that such an organization would serve to overlap and perhaps weaken the Congress. Surendranath Banerjea later on opined, 'I must say that the League served to create the first division in the Congress camp after the reunion. I did not join it, nor did many of the ex-Presidents of the Congress. I incurred some unpopularity ...

I had helped to build up the Congress. It was a part of my life work, my pride and my privilege, and it was not in me to weaken its influences or the great position which it occupied in the estimation of the country.'[36]

Under these circumstances, the moderates did not do anything by which the resolution of the Bombay Congress could be implemented. The draft Home Rule scheme which was to be prepared and approved by the All India Congress Committee before 1st September 1916, could not be produced by that date and now Mrs. Besant thought it essential and obligatory on her part to announce the Home Rule League and devise ways and means for its planning and organization, on a regular basis. It was, therefore, inaugurated in September 1915, and within a few days, its branches were formed in prominent towns of India, viz. Bombay, Madras, Kanpur, Allahabad, Banaras, Calicut and elsewhere.

'Her ideas in her talks stemmed partly from a series of articles which a leading British Conservative, Sir Valentine Chirol, had published in *The Times* in 1910, and which were expanded into his book, *Indian Unrest*. Here he paid unwilling tribute to the contributory influence of Theosophists from H.P. Blavatsky and H.S. Olcott to Annie Besant, laying special stress on Mrs. Besant's role in fostering a spirit of Indian nationalism. Women also had rights in India, and these too she preached in her lectures, without recommending that they go quite so far in their methods as their English sisters were doing. For many years she had been urging Indian women to participate openly in social life, including that of theheosophical Society.[37]

In a letter to *New India* Arundale stated that the Home Rule League had 110 branches and 5000 members in Madras Presidency.[38] He also inculcated the idea amongst the people that persons who desired Home Rule should pray for it everyday at noon.[39] In fact, the idea of gaining Home Rule by efforts of united will-power was earlier suggested by Tilak.[40] So in pursuance of this idea, a leaflet was printed at the Vasanta Press, containing the following exhortation.[41]

'Home Rule for India, use thought as well as speech and action'.[42] Translate into your mother-tongue and repeat aloud or to yourself everyday at noon the following sentence, '*India Shall Have Home Rule* and wish it with all your heart. Do not forget.'[43] On the four corners of the leaflet the following sentences were printed, 'Thoughts are things. Thought is the heart of action. What we think that we become. Strong thoughts mean resolute actions.'[44]

The Lucknow session of the Indian National Congress in 1916 was presided over by Babu Ambika Charan Mazumdar. It was altogether a unique gathering attended by 2301 delegates, besides a large number of visitors from various regions of India. It led to the fraternization of the Hindus and Muslims as well as for the formulation of a scheme of self-government. In fact, the requisite preliminaries of this concordot were gone through at a few joint conferences of the prominent members of the Congress and the Muslim League at Allahabad in April 1916 and in Calcutta in the following year. In working out the details of this agreement, the quotas of the Punjab and Bengal in respect of representation to the Muslims in the legislatures remained to be settled. But these issues were settled at Lucknow. The essentials of this unity were the subordination of the executive to the legislature though the latter itself was still to consist of a 115 nominated element. But no less important was the reunion of the two wings of the Congress which had been separated about a decade ago at the Surat session of the Congress. It was indeed a point of great political significance to see Tilak and Khaparde sitting side by side with other prominent leaders like Surendranath Banerjea and Rash Behari Ghose. In the same session were present the other leaders like Annie Besant, G.S. Arundale and B.P. Wadia with the banner of Home Rule in their hands. Besides Gandhi, Polak, M.A. Jinnah, the Raja of Mahmudabad, A. Rasul and Mazar-ul-Haq were also present. The Congress-League scheme which was approved by the Congress received the immediate approval of the Muslim League.[45]

The resolution dealing with self-government of India declared that having regard to the ancient civilization and the public spirit shown, His Majesty the King-Emperor should be pleased to issue a proclamation announcing that it was the aim and intention of British policy to confer self-government on India at an early date; that as a definite step in that direction the Congress-League scheme should be granted; and that in the reconstruction of the Empire, India should be lifted from the position of a dependency to that of an equal partner, in the Empire with the Self-Governing Dominions.[46]

Besides, the Lucknow Congress in a resolution viewed with alarm the extensive use made of the Defence of India Act and the Bengal Regulation III of 1818, and urged that in the application of the Defence of India Act, which was an emergency measure, the same principle should be followed as under the Defence of the Realm Act of the United Kingdom and the same procedure adopted in dealing with persons sought to be proceeded against under the Act that a statement in writing of the charge should be handed over to an arrested person, and in explanation obtained from him placed before a lawyer-judge of the High Court, a non-official practising lawyer, and a judge belonging to the Indian Civil Service, and no internment order before the same had been considered and the arrested persons afforded proper facilities for legal assistance.[47]

The same procedure was to be followed as far as practicable in regard to persons detained under Bengal Regulation III of 1818, and similar Regulations in other provinces. The movers of this resolution pointed out how extensive use had been made of the drastic provisions of the emergency measures. It was said that in Bengal about five hundred persons had been interned under Regulation III of 1818. Thus the rigorous enforcement of the Defence of India Act and the old Regulations had begun to be deeply resented by the people.[48]

In accordance with the terms of the resolution passed at the Lucknow Congress and under direction of All India Congress Committee, the Madras Provincial Committee appointed a body of twenty-five persons to carry out educative and propagandist work in connection with the scheme of self-government adopted by the Congress and by the Muslim League. Mrs. Besant, Arundale, Wadia and some other leaders of the Home Rule movement were placed on this body, and resolved among other things.[49]

(1) To prepare a brief exposition of the scheme of reforms adopted by the Congress and the Muslim League with a view to translate the same into Tamil, Telugu, Hindustani,

Kanarese and Malayalam, and spread its broadcast among the people.

(2) To send copies of the Congress scheme to all the non-official members of the Legislative Council, the members of the All India Congress Committee and leading citizens in the mufassal, requesting them to popularize the scheme by means of lectures in English and in the vernaculars.[50]

Mrs. Besant's political activity of less than a decade was marked by an indomitable will, concentrated purposefulness, undaunted courage and indefatigable zeal. 'Her superb oratory and matchless literary gifts enabled her to reach the foremost rank in politics in an incredibly short time. Even the moderates who detested her most, admitted that, 'she stirred the country by the spoken as well as written words as scarcely anyone else could do.'[51]

Undoubtedly her contribution to the widening horizon of the national movement was a significant phenomenon in Indian politics. Her message of Home Rule reached the remotest hut in some regions of India. She was responsible for introducing an agitational style for demanding *swaraj* for India. Her efforts to achieve unity among moderates and extremists and Hindus and Muslims were greatly acclaimed by one and all. She was quite clear in her mind that freedom did not descend to people, but was acquired by them. It was her ideology and unshakable faith to which she stuck like a rock.

Some well-known Indians of the time, like Gandhi, Jinnah, Tilak, Tagore and C.P. Ramaswami Aiyar greatly eulogized the services of Besant. Gandhi stated, 'Cultivate the great qualities of Dr. Besant, namely, firmness, simplicity, self-control etc. She is one of the greatest orators of the world, because she speaks what she believes and acts according to what she speaks... She has the courage of her convictions and always puts her words into action... Imitate her unflinching determination and simplicity of life... Obtain the same strength and indomitable will that she possesses which alone will bring *swaraj*. India is not fit for *swaraj* without these qualities. Remove India's chains and then alone will we achieve our goal... It is Dr. Besant who has awakened India from her deep slumber and I pray that she may live long to witness a free India'.[52]

REFERENCES

1. *Speeches of Gopal Krishna Gokhale* (Madras: Natesan), 1920, p. 1006.
2. Report of the 13th Session of the Indian National Congress held in 1897.
3. *Speeches of Gopal Krishna Gokhale, op. cit.*, p. 698.
4. *Ibid.*, pp. 708-09.
5. *Ibid.*
6. The Split at Surat in 1907 is an Historic example in this regard.
7. *The Indian Review*, November 1905, pp. 750-51.
8. Tarachand, *History of Freedom Movement in India*, Vol. III (New Delhi: Publications Division, Ministry of Information and Broadcasting, Government of India), 1972, p. 366.
9. *Minto Papers, Minto to Morley*, 21 March 1907.
10. *Ibid.*, 16 May 1907.
11. *Ibid.*, 21 March 1907.
12. Report of Lahore Session of the Indian National Congress, 1909.
13. Minto Papers, Minto to Morley, 24 July 1907.
14. *Ibid.*, 3 October 1907.
15. Report of the Indian National Congress, Madras, 1914.
16. Proceedings of the Indian National Congress Session 1914-15.
17. Majumdar, R.C., *History of the Freedom Movement in India*, Vol. II
18. Tahmankar, D.V., *Lokmanya Tilak* (London: George and Allen), 1956, p. 21.
19. Tarachand, *History of Freedom Movement in India*, Vol. III *op. cit*, p. 456.
20. *Ibid.*
21. *Ibid.*
22. *Ibid.*
23. *Debates*, House of Common, Vols., Vol. 66, Col. 955.
24. *Ibid.*, Vol. 68, Cols. 1357-58.
25. *Ibid.*
26. *Ibid.*
27. Raj Kumar, *Annie Besant's Rise to Power in Indian Politics 1914-1917* (Delhi: Concept Publishers), 1981, p. 134.
28. Besant, Annie, *Builder of New India*, Besant Spirit Series (Adyar: The Theosophical Publishing House), 1942, pp. 269-73.
29. *Ibid.*, pp. 272-273.
30. Home Deptt. Pol. A. October 1916, 36-53 & K.W.
31. Sitaramayya, B. Pattabhi, *History of the Indian National Congress 1885-1935* (New Delhi: Publications Division, Ministry of Information and Broadcasting, Government of India), 1972, pp. 220-221.
32. *New India*, 4 September 1916.

33. *Ibid.*
34. *Ibid.*
35. *Ibid.*
36. *Nation*, pp. 237-38.
37. Nethercot, Arthur H., *The Last Four Lives of Annie Besant* (London: Rupert Hart Davis), 1963, p. 29
38. Home Deptt. Pol. Dep., March 1917, File 32 & K.W.
39. *Ibid.*
40. *Ibid.*
41. *Ibid.*
42. *Ibid.*
43. *Ibid.*
44. *Ibid.*
45. Proceedings of the Lucknow Session of the Indian National Congress Session, 1916.
46. *Ibid.*
47. *Ibid.*
48. *Ibid.*
49. Home Deptt. Pol. Dep. April 1917, File 60.
50. *Ibid.*
51. Chintamani, C.Y., *Indian Politics Since the Mutiny* (Allahabad: Kitabistan) 1947, p. 102.
52. *New India*, 2 October 1918.

5

SPIRITUAL NATIONALISM

But why did Mrs. Besant come to India? Colonel H.S. Olcott, the founder of the Theosophical movement, explained in his *Old Diary Leaves* that she was sent to India by "the Rishis to fructify the seeds which had been planted by the Theosophical Society in India during the previous fifteen years".[1]

Indian Theosophists had invited her to visit India and to give lectures at the Adyar Convention of 1892. She could not come then, but she came in November 1893. From then on there was no respite. She toured villages, towns, and cities and was bestowed such honour as is usually given to "gods and persons of high spiritual renown".[2]

At one of the meetings held at Berhampore in West Bengal on 18 January, 1894, a thunderous reception was arranged in her honour:

"Welcome sister, the ever unfortunate mother India takes you to her bosom. Now she has nothing precious of which she can make a present to you; but she, is ready to receive you with *Shamit* (sacrificial fuel), *Kushahan* (a seat made of sacrificial grass), *Padya* (water for washing the feet with), *Arghya* (respectful oblation) and sweet words.

"What has brought you sister here? India is now lifeless. Here is now no chanting of the Vedas, no *Tapobana* (garden for practising religious austerities), no twice-born, no uttering of *Mantras* (Mystical incantation). Now the cry of famine-stricken people rends the sky.

"We, the inhabitants of Berhampore, give a garland of flowers round your neck: please take it, simple sister with your characteristic effability.

You are now a learned daughter of mother India, you are honoured throughout the world, and your reputation is world-wide. We are glad to see you."[3]

She felt as if India was her motherland in her earlier birth. Amidst warm feelings poured out to her, she realized for the first time what love of country meant! She felt as if she belonged to the people. She was happy that at last she had come to the place she loved and tried even blindly to serve before actually coming to India.[4]

To her, India was in very truth the Holy Land, the land whose great religion had been "the origin of all religions, the mother of spirituality, the cradle of civilisation".[5] She said that she thought of India in the greatness of her past, not in the degradation of her present. The India that she loved was the India whose "polity was built by *King-initiates* and whose religion was moulded by divine men".[6]

Her coming to India was thus prompted by her disenchantment with the material values of life. Through her studies and reflections she was by now thoroughly impressed by the great religion and philosophy of India. She had also understood by her theosophical learning that there existed higher nature in man. She believed that it was in the revival of this quality in man that the safety of the world could be ensured, and true service to the humanity rendered. She claimed to have found the truth of life. She became impatient to try her newly found idealism of establishing the "Brotherhood of Man". She considered India to be the right place for such a task. She thus came to India in 1893.

For an analysis of her motivation in Indian politics, we may now study the Indian scene and her reaction to it. On her arrival, Annie Besant found India involved in a strange and serious conflict. Some people were turning mad for everything Western, while others felt disturbed about the eclipse of their customs and traditions. In the midst of all this the rulers were trying to establish a "Police State", where maintenance of law and order was the primary concern, and the well-being of the people just a promise.

The generality of the people welcomed the strong rule of the British because the consolidation of the British power in India was preceded by a period of internal disturbance and the general insecurity of life and property. British rule was also popular with various grades and classes of society who formed themselves into strong and well-entrenched vested interests. These were the English educated classes who got advantageous positions in new professions that were created by the British.

These Western educated classes started doubting the desirability of various social institutions that had grown up in the land throughout the ages. They even questioned the truth of many of the tenets of ancient faiths. These persons thought seriously that nothing "but complete metamorphosis of India, making her as near a copy as possible of the Western World, could save India".[7]

Still there were those who reversed their own traditions and customs. They were not influenced by the foreign lore. They endured to preserve their own age-old notions of right and wrong. They withdrew themselves into a shell in order to keep themselves at least unpolluted. They felt upset at the spectacle. They tried to preserve the sanctity of their faith and customs in their own persons as best as they possibly could.[8]

At this stage Annie Besant came to India. As a first step she learnt up the ancient lore of the land—its philosophy, its religion and the basic ideals of its culture'. She acquainted herself with the manners and customs of the people so that she might not remain a stranger. Then under the auspices of the Theosophical Society she travelled throughout the country. She delivered several lectures depicting in vivid language the glory and the greatness of ancient Indian thought.[9]

Mrs. Besant tried to tackle the situation in her own style. She lauded Indian religion and philosophy during her numerous lectures and speeches, and exhorted people to arrest the tide of Westernization and to encourage those who still had faith in their own customs and traditions. She actually lived and practised Hindu customs and ideals in her daily life. In 1898, she opened the Banaras College. She established a network of schools and colleges in the land. For this she did not accept financial assistance from the Government.[10] She never hesitated to admit such students as were turned out of the official colleges for having kept the picture of a Bengal revolutionary in their rooms or for similar other reasons.

Thus her lectures on ancient Indian thought and philosophy and her schools and colleges which popularized education and social reforms formed a bridge between the two sections of India's society that were drifting apart.

The partition of Bengal in 1905 created a ferment in the country. It gave an impetus to Swadeshi movement. Annie Besant had realized early the value of Swadeshi for the economic regeneration of India, and she was always pleading for Swadeshi. She herself wore India-made cloth. She eulogized *Swadeshi*, and criticized the partition of Bengal in the strongest terms.[11] But she was against students' participation in political movements. She considered politics, which affected millions of lives, to be the work of grown-up and mature minds. She prevented her students from participating in politics. As a result she became unpopular for some time, and was severely attacked in the Press. It was taken for granted that "her true colours were now revealed"; that "she was at best a representative of British imperialism"; that "all her work was a mere pretence".[12]

The unrest in India during 1905-10 affected the student community also. She organized such societies as would create a sense of discipline and give direction to the energies of the students for the good of the country. One such organization that she mothered was "Sons of India Society". This order consisted of men and women of all ages. The elders, by sympathy and good counsel, guided the energies of the youth into channels conducive to the progress of the country, and thereby enduring help to them to that discipline and self-sacrifice which made the citizens worthy to be free.[13] This, it was believed, would result in the "gradual widening into ordered freedom in which Government and the people shall alike rejoice."[14] In her endeavours she sought the blessings of the then Viceroy, Lord Minto, who thought that her idea of "Sons of India Organisation" was sound.[15]

While appreciating the good points in the British policy in India, she criticized it wherever she found it wanting. She held that the Minto-Morley Reforms were a step in the right direction.[16] She was also in favour of the 1858 proclamation that colour should be no bar to the service of the State.[17] India, she said, could not be contended while her sons were debarred from all the higher posts of administration, and that Indians were aware that England "withheld from them the very privileges and institutions she advocated for Persia, Turkey and Russia".[18] She disliked the treatment accorded to Indians in their own land, which, she felt, was not only unbecoming of the English race, but also was fraught with serious consequences, and she denounced it.

On 19 and 20 January, 1910, the Central Hindu College held its anniversary. One of the ex-students complained to Mrs. Besant that he had been maltreated by an Englishman on the way. She felt that due to such unworthy actions, the good name of the English was sullied, and embitterment created. She issued "An Appeal" to the Government and to Europeans:

> "We who have charge of nearly one thousand lads in the Central Hindu College, and who influence tens and thousands all over India, we who are straining every nerve to sow in these young hearts love of the motherland and of the empire, we who seek to win them to love England by making the English lovable, who love India, and hope to see her a self-governing part of this mighty Empire, we appeal to the Government of India not to allow this work of love and service to be wrecked by brutality and folly."[19]

The Government did not relish it and proposed to take action against her. She wrote to Gopal Krishan Gokhale about it. Gokhale consulted S.P. Sinha and met officials in the Home Department. The Government decided to drop the matter,[20] but the officials had taken a serious view of it. Lord Minto, the Viceroy, while agreeing with the views expressed by the officials noted that he had no doubt at all that Mrs. Besant was perfectly honest in her intentions and "he knew privately that she was very unhappy at the interpretation given to her Appeal".[21]

Mrs. Besant's activities in the Central Hindu College magazine came under severe criticism of the officials. H.V. Lovett, the Commissioner of Benaras, reported that she was entirely unworthy of the confidence with which she had been treated. He added that she should no longer be allowed to use the college magazine for her own political propaganda which was calculated to do mischief in the "extreme" when addressed to youth, and was not the creed of at least some of those whose name and money she used.[22]

Bhagwan Das, in a letter to Lovett on 20 February, 1910, expressed his regret for misunderstanding on the part of the local government about the Open Appeal by Mrs. Besant. He explained that her appeal was for the promotion of goodwill and mutual trust between Englishmen and Indians.[23]

Bhagwan Das added that "all that was in the spirit of the King and his wise minister Lord Morley's words, that class for class, education for education, character for character, Englishmen and Indians shall be equal subjects." He pointed out that it was just this task that Mrs. Besant and her colleagues had been trying to perform and that it was saddening that instead of being helped by the local government, she was completely misrepresented.[24]

Mrs. Besant's activities were being watched closely and reported regularly. On 2 April, 1910, Lord Minto agreed that information contained in Mr. Lovett's note was disturbing, and that the Central Hindu College required careful watching. Although he reiterated his faith in her loyalty, yet he noted, "she is an enthusiast who is carried away by her own ideas which are sometimes positively dangerous".[25] Again on 23 May, 1910, the Government records reveal that it was considered dangerous to "give any countenance to Mrs. Besant in any enterprise which she might undertake". She might have good intentions, but she was of unstable mind, and it was difficult to anticipate what she might do or say from one day to another. "I believe she finds it occasionally".[26]

Mrs. Besant did not care for the 'reaction of the Government towards her activities. On 27 July, 1910, she sent a petition for a Royal Charter for the University of India to the Viceroy. This was signed by all sections of the Indian people. The Government dubbed her ideal of establishment of the University of India reactionary and politically dangerous.[27]

She believed that the English rule under the Company reduced India to poverty, and almost destroyed her industries and that under the Crown the Government still hampered her industries, made a cruelly severe drain upon the country, and by their fiscal arrangements prevented the return of prosperity.[28]

Mrs. Besant could no longer stick to her resolve of not resuming any political work as increasing repression by the authorities—the curtailment of liberty, the ill-treatment of students, and the danger of revolution—forced her into the field. It had become difficult to do any educational, religious or social work, and the boys she loved were being thrown into crime. Liberty was strangled and she wrote, "I as one of her old soldiers could not stand aside. I joined the political campaign not to lead but to take risk".[29]

The Indian National Congress which was a loyal organization was also losing faith in England's love of liberty in the face of the Arms Act, the Press Act, the house searchings, the espionage, the autocracy, and the frustration of the hopes of people of India.

In these circumstances, the relations between the Indians and English people became strained. But she longed to see India and Britain together as guardians of the peace of the world. She had a clear understanding that British rule in India while efficient in less important matters and those which concerned British interests was inefficient in greater matters on which the healthy life and happiness of the people depended.

Thus Mrs. Besant never denied India's right to liberty and independence. But she strongly believed that for the good of England and India and for the prosperity and happiness of the whole world, it was essential to strengthen the ties between the two countries, of course, on the basis of brotherhood and perfect equality.

But as the things were, she became convinced that nothing but a change in the Government of the country could make her work for the uplift of the people possible.[30]

Was it then the sufferings and oppressions of Indians which evoked in her the earlier political agitator for similar causes in England and elsewhere? Her colleague Ramaswamy Aiyar attributed her entry into Indian politics to altruistic interests and to her passionate desire for the elevation of humanity.[31] Was it her ambition to assume the leadership of the Nationalist Movement as her biographer Mrs. William assessed it?[32] Was it for fostering Indo-British friendship, and a wish to march as the head of the Congress to self-government as H.F. Owen has analysed?[33] Was it under instructions received by her from the Master known as "The Regent of India" or to serve as a safety-valve as others believe?

The nature of man has in it certain sentiments and emotions which, reasonably or unreasonably, sway him powerfully and continually; they are in fact his strongest motive powers, overwhelming the reasoning faculties with relentless strength; true they need discipline and controlling, but they do not need to be, and they cannot be destroyed. The sentiments of love, of reverence, worship are not as yet reducible to logical processes; they are institutions, spontaneous emotions, incomprehensible to the keen and cold intellect. They may be laughed at or denied, they still exist in spite of all; they avenge themselves when they are not taken into account by ruining the best formed plans, and they are continually bursting the cords with which reason tries to tic them down.

A probe into the foregoing analysis and an insight into the motivational factors of human actions reveals that Annie Besant's actual motive seems to have eluded the grasp of the scholars in the field. Although the validity of any of the above motivational factors may not be ruled out, yet the issue does not appear to have been clinched.

Man is born into a world which swaddles him in the material and spiritual inheritance of society. Purely material and immediate success does not suffice, and that mind has to be satisfied as much as the interest.

The epitaph that she asked for on her tomb was: "She tried to follow truth". It was in this spirit that she moved from one field of activity to another, experimenting with life from different points of view but always in the spirit of ennobling adventure, and it was this fund of spiritual strength that was in common with Mahatma Gandhi. After sampling many movements, she was troubled as to the "beyond" of all her efforts of social and political reforms. She longed for the motive which would enable her to establish a Brotherhood of man.

Feelings of deprivation create high tension seeking an outlet. Those who are compensating for intense feelings of deprivation have been found to be important leaders. An analysis of Annie

Besant's motivations gives enough evidence to support this contention.

Annie Besant was a highly religious, emotional, and proud individual, who had a frustrated and almost broken married life. She underwent cruelty and suffering through her husband and the illness of her children. She even thought of ending her life. For finding an anchorage amid the stormy sea of her conflicts, she studied, reflected and wrote. She turned a free thinker, drifted to Fabian Socialism but found no true happiness for herself and for the world around her. Then like other enlightened beings she wandered in search of truth. She caught a glimpse of reality in theosophy. Herein she got the proof of existence of a higher nature in man. India, she believed, was the home of such a higher power in man.

On arrival in India she saw that the "un-British" rule by the bureaucracy in India was dwarfing the Indian character. Through religious, educational and social works, she tried to restore among the people faith in their own great religion and customs. She found the bureaucracy cold and suspicious towards her efforts. Due to the conditions created by the Government it became difficult for her to proceed with her work. She, therefore, realized that a change in government alone could make her work possible, and she ultimately decided to plunge into the whirlwind of active Indian politics in 1914.

The Spiritual Theory of Nationalism

Annie Besant had a sustained and deep love for India. In 1930, she wrote a poem calling India to rise up:

> "O India ! O perfect Nation!
> O India that shall be !
> How long till Thou take station?
> How long ere thralls live free?
> How long ere all Thy Soul
> Be one with all Thy sea?"[34]

She repudiated the materialist approach to nationalism which conceives of it as merely an epiphenomenon of capitalism. To her the nation was a spiritual entity pulsating with a deep inner life.[35] She traced the roots of Indian nationality in the ancient literature of India and in the past embodied in it. In her presidential speech at Calcutta Congress Besant said:

"For, what is a Nation ? It is a spark of the Divine Fire, a fragment of the Divine Life, outbreathed into the world; and gathering round itself a mass of individuals, men, women and children, whom it binds together into one. Its qualities, its powers, in a word, its types depend on the fragment of the Divine Life embodied in it...The magic of Nationality is the feeling of oneness, and the use of nationality is to serve the world, in the particular way for which its type fits it. This is what Mazzini called 'its special mission', the duty given to it by God in its birth-hour. Thus India had the duty of spreading the idea of Dharma, Persia that of Purity, Egypt that of Science, Greece that of Beauty, Rome that of Law. But to render its full service of Humanity it must develop along its own lines, and be self-determined in its evolution. It must be itself, and not another. The whole world suffers where a Nationality is distorted or suppressed before its mission to the world is accomplished."

She regarded the nation as the vehicle of divine manifestation. Thus, like Fichte, Hegel and Aurobindo, she gave a spiritual theory of nationalism.

The Nation, according to Besant, is a spiritual entity and an unique revelation of God. Every nation expresses some essential truth of the divine being. This unique external expression is symbolized in the national character of a people. She wrote:

"What is it...that makes a Nation ? It is, as with other Individuals, a fragment of God Himself, a *Jivatam*, a living Self, with innate qualities which gradually appear and form its Character. Contrast the Indian and the Englishman, and you will see the difference in the National Character: the Indian, spiritual, courteous, polished, keenly intellectual, inclined to philosophy and poetry, with an acute sense of duty, of obligation, to his surroundings; the Englishman, somewhat blunt and abrupt, strong mentality, scientific and practical, public-spirited. Climate, environment, social customs, all act on physical peculiarities, and through them on character also. A Nation is distinctly an Individual with a Character, and that character depends on the nature of the Spirit at its core, and its gradual unfolding to play its part in humanity as a whole....If India still survives all who were her contemporaries five thousand years ago, it is because the same Spirit lives in her national body as lived in it then."[36]

Thus, according to Besant, the nation is in individual and a spiritual entity. Like Hegel and Aurobindo and Pal, Besant also regards the nation as the unfoldment of the supreme divine. National religion, according to her, can be a factor for the survival of the nation even when it has its own government and territory destroyed, as happened in the case of the Hebrews.[37] A nation becomes a state when a fixed territory and government are added to it.[38]

Besant did not believe, that India had to learn to be a nation, from the west. India was and is a nation. She had been a nation from the past. Throughout her literature, her philosophy, her art, there is a deeply pervasive current of vibrant national spirit. Many civilizations rose in the world but eventually they collapsed. But India's fidelity to the religious roots of nationalism kept her vital and in a position to regain her lost energies. To say that the Indian Feeling of nationality has grown out of British rule is "silly" and "absurd".[39] It is superficial to stress that nationhood depends on the unity of language or the unity of race. Nationhood is a spiritual phenomenon. Its essence of national vitality and perfection lay not in identity of opinion but in identity of aspiration.[40] Where a large body of men were imbued with a high common purpose, national unity was bound to come. Besant wrote:

"....a Nation, like a man, is the building up of a complicated body which is inhabited by a higher life—m—A life that is divine. Just as each one of you is a *Jivatma*—a portion of the Self of *Ishvara*—that moulds your character, shapes your destiny, and is an impulse for your evolution, so it is with the life of a Nation : a Nation is an individual of a higher order. The spirit of a Nation is a fragment of *Ishvara,* comes direct from God Himself and, according to the peculiarities embodied in that fragment, are the characteristics of the Nation which is built up thereby. No two individuals are alike and similarly, no two Nations are alike. It is the totality of all Nations that builds up Humanity—the human reflection of *Ishvara* Himself. Each has its own individuality. That which makes the richness of the national life of India is its marvelous literature, its history, its religion, its science—and all these are so highly evolved because the Nation is so ancient a Nation. During the early life of a Nation, religion is essential for the binding together of the individuals who make the Nation. India was born, as it were, in the womb of Hinduism, and her body was far long shaped by that religion. Religion is a binding force, and India has had a longer binding together by religion than any other Nation in the world, as she is the oldest of the living Nations."[41]

Besant believed that Indian spirituality was to be the saviour of the world. That was the great destiny of this country. Thus, like Vivekananda and Aurobindo, Besant also believed that India had spiritual 'mission' for the world.

Besant advocated the politically revivifying force of a religious synthesis. Religion teaches solidarity. The great religions of the world have contributed to the moral enhancement of the human consciousness and to the enrichment of the cultural heritage. India could be a beacon light to the world if these religious sources of strength could be canalized. Besant wrote:

"I feel strongly .... that only on the basis of religion can a true Nationality be built up—Religion itself which gives the sense of human dignity joined to the sense of human duty, and the self-respect of India could never have arisen as it has arisen today, had not the philosophies and the religions of past re-won their empire over the Indian heart; and the virtues that the great religions typify, all of which we find in this sacred land, are virtues that we need for the building of our Nation. Shall we not learn from Hinduism to see the one Self in all, and to know that in men of every caste, of every class to see the Self equally dwelling is to touch the central truth of that most ancient faith? Shall we not learn from Zoroastrianism the necessity for National purity, from Buddhism and Jainism the need for wisdom and right thinking ? Shall we not learn from Islam that great lesson of true democracy which Islam above all other religions of the world exemplified in the life and teachings of the great Prophet? Shall we not add to these the courage of the Sikhs and so make perfect the virtues of a great National life ? And shall we not realize that Christianity brings also the great jewel of sacrifice and its lesson? Thus... looking at all the religions of the world as rays of the one Light, these religions shall join together to make the white light of India a Nation, nothing left out, nothing excluded, learning from each other and trying to admire and serve each other."[42]

Besant had a glorious ideal of India's destiny. She had the vision of an Indo-British Commonwealth and a World Commonwealth.[43] She believed in the 'Brotherhood of Humanity', and in her presidential speech at the Calcutta Congress in 1977, she thundered:

"To see India free, to see her hold up her head among the Nations, to see her sons and daughters respected everywhere, to see her worthy of her mighty Past, engaged in building a yet mightier Future—is not this worth working for, worth suffering for, worth living and worth dying for? Is there any other hand which evokes such love for her spirituality, such admiration for her literature, such homage for valour, as this glorious Mother of Nations from whose womb went forth the races that now, in Europe and America, are leading the world? And has any land suffered as our India has suffered since her sword was broken on Kurukshetra, and the peoples of Europe and of Asia swept across her borders, laid waste her cities, and discrowned her Kings. They came to conquer, but they remained to be absorbed. At last, out of those mingled peoples, the Divine Artificer has welded a Nation, compact not only of her own virtues, but also of those her foes had brought to her, and gradually eliminating the vices which they had also brought. After a history of millennia...India, who has been verily the Crucified among the Nations, now stands on this her Resurrection morning, the Immortal, the Glorious, the Ever-young; and India shall soon be seen, proud and self-reliant, strong and free, the radiant Splendour of Asia, as the Light and the Benediction of the World."

India throughout the ages had stood for justice, duty, fitness and right order. Hence India should first come to the realization of her own destiny and then lay her legitimate role as the temple of humanity. This was the divine plan and for this the great Mahatmas and teachers were working.

A Commonwealth of Brotherhood

Nationalism is a spiritual phenomenon. It is the expression of the inner soul of a people. A nation is the manifestation of the divine being. But nationalism is only a process, a stage, and not the culmination of social evolution. It is to receive its fulfilment in the realization of a world brotherhood. Like Mazzini, Gandhi and Aurobindo, Besant with all her eloquent rhetoric in the exaltation of the nation, accepted it as a stage—the stage of individuality. The second or higher stage was the consummation of a commonwealth of cosmopolitanism. Besant wrote:

"The next stage of the Plan is the building up of a free Commonwealth of Nations in which India plays her equal part. This is why the English came here and others had to go away, and, the one Nation that is free in her institutions in her own island, is not free in her practice outside her own island, is the English Nation, chosen to come here and lend with the Indian Nation in the building up now of the World Empire, to be really a World Commonwealth, not a World Empire which was to rule by force but a World Federation which was to govern by love and in peace. That is the ideal towards which some of us are trying to work, towards which the Manu is working and He asks for co-operation from His elder children here, to bring together East and West…harmoniously blending them for helping of future generations with the great spiritual ideals of India and the great material and scientific progress of Britain. These two are to be the main constituents of this commonwealth, which is to be the model of the World Commonwealth of the future. It is a model on a small scale of internationalism—that is the great building which Vaivasvata Manu is trying to bring about, against the clashing Wills of men and against all strivings of the ignorant and, far worse against those Powers of Darkness who are ever the opponents of the Brothers of Light."[44]

Besant was confident that Britain would play her part, in this plan and would thus vindicate the supremacy of justice. She wrote: "Britain's opportunity is hers, because all round the world there are Free Nations that have sprung from her, that you call the Self-governing Dominions, and other lands that have been acquired largely by the help of their own people, and which you call Dependencies or Colonies; they...are awaiting federation.

Think what it means if, for the first time in the world, a Power as strong as you admittedly are today, instead of relying on strength, tries to seek for and to do justice. Instead of trying to tyrannize over others, open to them the gates of Freedom, and say to all the Nations that make up this great Empire: 'Come and form with us not an Empire, but a great Commonwealth of Free Nations; not a White Commonwealth, but a Commonwealth into which men of every race, of every colour, of every ancestry, of every creed, of every tradition and custom, shall come willingly as free members.' Ah ! If Britain can do that, then she will do her part in the Great Plan. That is her place; that is her opportunity."

Besant had great hopes that East and the West, Asia and Europe would march together as equals and would thus help humanity.

Besant's concept of universalism was based on an organic view of the individual, society, nation and humanity. She tried to harmonize her concept of spiritual immanence with the concepts of Bluntschli and Herbert Spencer, regarding organicism. She categorized seven grades of individuals:

1. Cellular lives.
2. Cells joined together into Tissues.
3. Tissues joined together into Organs.
4. Organs joined together into Bodies.
5. Bodies joined together into Communities.
6. Communities joined together into Nations.
7. Nations joined together into Humanity.

In her Lectures on Political Science, Annie Besant recognized eight categories of biological individuals: (i) A simple cell, (ii) A group of cells forming an organism, (iii) A simple or complex organism up to a human being, (iv) Group of human beings forming a family, (v) A group of families forming a tribe, (vi) A group of tribes forming a nation, (vi) A group of nations forming an empire or commonwealth, and (viii) A group of commonwealth or empires forming organized humanity. The Divine being unfolds itself in each successive form.

The Political Ideas of Besant

(a) ***Liberty***. In the life as well as in the teachings of Besant there is a great quest of liberty. In her early life she had broken from the restrictions of the Church of England and joined the Free Thought Movement of Charles Bradlaugh because she had an earnest quest for freedom of conscience. She also claimed to have started the 'Home Rule for India' idea while she was still in England in 1877. She wrote:

"I demand for everyman, whatever be his opinions, that he shall be able to speak out with honest frankness the results of honest thought, without forfeiting his rights as citizen, without destroying his social position, and without troubling his domestic peace .... Liberty is immortal and eternal, and her triumph is sure, however, it may be delayed ... and hers is also—the triumph of tomorrow if only we who adore her, if only we can [be] true to ourselves and [to] each other. But they who love her must work for her, as well as worship her, for labour is the only prayer to Liberty, and devotion the only praise."[45]

Thus although, like Aurobindo, Besant would also conceive of liberty as an eternal attribute of the soul, she stated that freedom was a precious heritage which had to be won by great efforts and discipline. Freedom is the farthest removed from arbitrariness and licence. It is based on positive fulfilment through the conservation of moral and spiritual energies. Hence the inner freedom of spirit is essential to be cultivated before freedom could be attained in the external sphere. The subjugation of the appetites of the empirical ego alone could provide the discipline and character essential for the realization of freedom.[46] Besant wrote:

"Liberty is a great celestial Goddess, strong beneficent, and austere, and she can never descend upon a nation by the shouting of crowds, nor by arguments of unbridled passion, nor by the hatred of class against class. Liberty will never descend upon earth in outer matters until she has first descended into the hearts of men, and until the higher spirit which is free has dominated the lower nature, the nature of passions and strong desires, and the will to hold for oneself and to trample upon others. You can only have a free nation. when you have free men to build it out of free men and women both; but no man is free and no woman is free who is under the dominance of appetite, or vice, or drunkenness, or any form of evil which he is unable to control. Self-control is the foundation on which alone freedom can be built. Without that you have anarchy, not freedom, and every increase of the present anarchy is paid for by the price of happiness, which is given in exchange. But when freedom comes, she will come down to a nation in which every man and every woman will have learned self-control and self-mastery; and then, only out of such men who are free, out of such women who are free, strong, righteous, ruling their own nature and training it to the noblest end—of such only can you build up political freedom, which is the result of the freedom of the individual, and not the outcome of the warring passions of men."[47]

For the realization of freedom, it is essential to conform to Dharma based on the relationship of all sentient creatures to one supreme divine being. Hence Besant pleaded for the renunciation of the concept or man as a solitary individual clothed with rights in isolation. In her speech at the Madras Congress in 1914 she referred to Milton and Mill in support of liberty. At the Bombay Congress of 1915 she characterized the Regulation III of 1818 as a "shameful revival of the old Bourbon barbarism."

(b) *Aristocratic Socialism.* Besant had begun her public life as a socialist. She was an opponent of the combativeness of individualism and preached the virtue of associative co-operation. But she could not sympathize with the upsurge of mass emotions and the rigid doctrinaire equalitarianism that have been associated with socialism.[48] She wanted a socialism based on common agreement leading to the realization of "wisely adapted and mutually beneficial and enjoyable adjustment of capacities and functions." Instead of mass socialism, she pleaded for the rights of the aged and the learned to control the governmental process. Thus she had a Platonic approach to the problem of sovereignty according to which rulership belongs not to the unregenerate numbers but to those who are intellectually and, morally trained and disciplined. She wrote:

"....we should restore to the State something of the knowledge that it wants, and we should take away from the State the danger of allowing a mass of ignorant electors—who are really fighting to elect a man who will look after their mines, their drains, their local interest, matters they understand to upset international arrangements, and possibly plunge us into war—or worse, into dishonour. Those ire the general principles which might be worked out, and might be applied to modern days...A democratic Socialism, controlled by votes, guided by numbers, can never succeed; a truly aristocratic Socialism controlled by duty, guided by wisdom, is the next step upwards in civilization."[49]

But the aristocracy which, Besant had in mind is not a plutocratic elite, but the aristocracy of wisdom and moral power where authority is to belong to righteousness and wisdom.

(c) *Critique of Representative Democracy*. Besant had a Platonic philosophy of politics because she believed, not in the sovereignty of numbers but in the omnipotence of wisdom. If the doctor and the lawyer need special competence, there is no reason to ignore special skill in the voter who choose men to manage the nation's business.[50] She was painfully conscious of the fact that under the mask of external democratic structure many countries in the West were presenting the spectacle of chaos, ignorance and organized force, and hence she criticized the practice of mere counting of heads and not weighing the contents of the heads.[51] She even referred to the dominance of "multi-headed ignorance" in the western democracies.[52] She was in favour of the right of spiritual and moral wisdom to wield the sceptre of punishment and wanted that the "wise" should be placed in seats of power.[53] Majoritarianism and the dominance of numbers can have but one consequence, the mutual clash of forces leading to 'chaos,' confusion and the triumph of cunning manipulators. The sole remedy is that the wise men should be entrusted with the function of rulership. Only those who are unselfish, resolute in their determination to subserve the common good and wise, should take upon themselves the burden of governance. To them a governmental position is an opportunity not for personal aggrandisement but of social service. She wrote:

".....our Ideal of Brotherhood applied to Government, claims power for the wise and not for the ignorant; places law-making who understand the complex problems of industry, and not in the hands of those who only know the needs of the household, or perhaps the needs of the town. The people have a right to happiness, but they can never get it for themselves to be guided to it by knowledge of those who understand... When organisation is unselfish, instead of selfish, then this labour problem will be solved and not till then. Now how to find the best? What is that problem? Everyone of us who studies must try to solve this problem... But you will not solve it until you realize the hopelessness of the present line of ruling—or not ruling—and accept the ideal that the best should govern... There lie much of the work of Theosophy where it is dealing with the problems of Government. Oppose all that would level down, but help all that would level up. Do not allow the culture and refinement of ages, gathered by toil and pain, by the long struggles of generations, to be overwhelmed in one mass of ruin by the ignorant and the foolish, as has happened so many times before....To those problems of Government.....bring the power of a great Ideal."[54]

But in the context of the advance in self-government, constitutionalism and democracy in India and Asia in the twentieth century, Besant's conception of the sovereignty of the moral aristocrat may appear anachronistic. It will be difficult to convince the common man of his incompetence. It is impossible that the Indian masses would willingly give up the grant of adult suffrage to them. In the eyes of Indian nationalists, Besant's theory of educational and moral aristocracy would thus appear to be a subtle plan to defend the policy of the British government to restrict the suffrage as was done in the Indian Councils Act of 1909 and the Government of India Act of 1919. But Annie Besant although an exponent of Platonic ideals of government, was very strangely in favour of the village panchayat which she regarded as the true democracy of the East and pleaded for its rebuilding.[55]

REFERENCES

1. Olcott, H.S., *Old Diary Leaves* (Adyar: Theosophical Publishing House), 1975, p. 92.
2. Nethercot, Arthur H., *The Last Four Lives of Annie Besant, op. cit.*, p. 17.
3. Olcott, H.S., *Old Diary Leaves, op. cit.*, p. 101.
4. *The Besant Spirit*, Volume XI (Adyar: Theosophical Publishing House), 1943, p. 19.
5. *Ibid.*, p. 18.
6. *Ibid.*
7. Sri Prakasa, *Annie Besant* (Bombay: Bharatiya Vidya Bhavan), 1954, p. xx.
8. *Ibid.*, p. xxi
9. *Ibid.*, p. 85.
10. *Theosophical Society Archives* (Adyar: Theosophical Publishing House), No. B 111/20, p. 3.
11. *Home Poll*, Deposit, May 1910, No. 10.
12. Sri Prakasa, *Annie Besant, op. cit.*, p. 75.
13. *The Statesman*, 22 October, 1908.
14. *The Tribune*, 6 November, 1908
15. *Home Poll*, Deposit, February 1909, No. 25.
16. *Ibid.*, May 1910, No. 10.
17. *Ibid.*
18. *Ibid.*
19. *Ibid.*
20. Gokhale to Annie Besant, 10 March 1910, Annie Besant Papers, T. S. Archives, *op. cit.*
21. *Home Poll*, Deposit, May 1910, No. 10.
22. *Ibid.*
23. *Ibid.*
24. *Ibid.*
25. *Ibid.*(Note by Minto dated 2-4-1910)
26. *Ibid.* (Notes by J. L. Jenkins dated 23-5-1910)
27. Lutt Jurgen, "The Movement for the Foundation of the Benares Hindu University, German Scholars on India", Contributed to Indian Studies (ed.) by the Cultural Department of the Embassy of Federal Republic of Germany, Vol. 2, New Delhi, 1976, pp. 160-95.
28. Besant, Annie, *A Bird's Eye View of India's Past as the Foundation for India's Future* (Adyar: Theosophical Publishing House), 1931.
29. *New India*, 4 April, 1917.
30. Interview with Montague, Theosophical Society Archives No. B111/20, p. 2.
31. Aiyer, C.P. Ramaswamy, *Annie Besant, op. cit.*, p. 18.
32. Williams, Gertrude Mervin, *The Passionate Pilgrim* (New York), 1931.
33. Owen, H.P., "The Home Rule Leagues, 1915-18", in Low, D. A. (ed.), *Soundings in Modern South Asia History* (London: Oxford University Press), 1968, pp. 165-66.

34. Besant, Annie, "O India! Awake! Awake!", *New India*, May 1, 1930.
35. Besant, Annie, *How India Wrought For Freedom* (Adyar: Theosophical Publishing House), 1915, p. ii.
36. Besant, Annie, *Lectures on Political Science* (Adyar: Theosophical Publishing House), 1918.
37. *Ibid.*, p. 33.
38. *Ibid.*, p. 69.
39. Besant, Annie, *Shall India Live or Die*? (Adyar: Theosophical Publishing House), 1925, p. 38.
40. *Annie Besant*, New India, April 16, 1918.
41. *Ibid.*, February 21, 1917.
42. *Ibid.*, September 27, 1917.
43. Besant, Annie, *The Future of Indian Politics* (Adyar: Madras), 1922, pp. 314-15.
44. Besant, Annie, *The Great Plan* (Adyar: Theosophical Publishing House), 1920.
45. Besant, Annie, *Civil and Religious Liberty* (Adyar: Theosophical Publishing House), 1883.
46. Besant, Annie, *Popular Lectures on Theosophy* (Adyar: Theosophical Publishing House), 1883, p. 133.
47. Besant, Annie, *The Changing World* (Adyar: Theosophical Publishing House), 1909.
48. Besant, Annie, *Lectures on Political Science, op. cit.*, p. 133.
49. *Ibid.*
50. *New India*, April 21, 1922.
51. Besant, Annie, "Shall India Live or Die? ,*op. cit.*, p. 152.
52. Besant, Annie, *The Future of Indian Politics, op. cit.*, pp. 275-76.
53. *Ibid.*, p. 215.
54. Besant, Annie, *The Ideals of Future of Theosophy* (Adyar: Madras), 1912.
55. *Ibid.*, p. 113.

6

MRS. BESANT, GANDHI AND CONGRESS

Outside her Theosophical preoccupations, which she maintained during the war in routine fashion, Mrs. Besant's two chief interests had been education and politics. The two were now to come in collision in an episode which was further to condition her relations with the man who was to play perhaps the greatest role in her political future, Mohandas Gandhi. After what seemed like years of negotiating, compromising, and often downright bickering, plans for a magnificent foundation stone laying of the new Hindu University were formulated for a four-day ceremony, beginning on 3rd February 1916. The event had been publicized all over India, when Mrs. Besant spoke to Pandit Malaviya at an open-air money-raising meeting in Madras early in January. So much national enthusiasm had been worked up that thirty special trains had been chartered to bring the visitors and the famous guests. It was to be full-dress affair, with the retiring Viceroy and all kinds of maharajas, rajas, and other titled personages, Indian and British, to be on the platform. The Maharaja of Dharbhanga, as President of the University Committee, was to read the opening address. There were to be parades, cricket matches, songs, and dramatic performances by the students. And of course speeches upon speeches were scheduled, the culminating one of which, on "The University as a Builder of Character", was to be delivered by Annie Besant, as founder of the original Central Hindu College. Mohandas K. Gandhi, whose influence on Indian students was growing everyday, was to precede her.

Everything went according to plan until the last night. Whether the fact that troops and CID men had been stationed everywhere, in the buildings and gardens, along the roads and even the railway line, to protect Hardinge and his officials from possible violence, was responsible for the good order and good manners of the crowd is problematical. At any rate, this sort of interference was resented as an insult by many Indians and regarded as a downright provocation by Annie Besant. Moreover, Gandhi had scheduled another speech that same evening in another place, to be delivered before he came to the University. Something delayed him, and Annie, who had expected the place of honour, had to go on first to keep the audience occupied. At last he arrived, and launched, more or less extemporaneously, into his speech. Exactly what happened next will never be known, since there were no tape-recorders in 1916, and the accounts of the various eye-and-ear witnesses contradict one another at almost every important point.

All that is certain is that the meeting broke up prematurely in utter confusion.[1]

Seen through the perhaps somewhat prejudiced eyes of Annie Besant herself, here is what happened at what V.K. Krishna Menon many years later referred to as the "Benaras Affairs". The Maharaja of Darbhanga, as presiding officer, introduced Gandhi, clad in his native Kathiawar costume, which in its very humbleness stood out against the rich clothing of the native princes and rajas seated on the opposite side of the platform from Mrs Besant. Speaking in his well-known quiet colourless voice, which was described as deliberate yet rasping by young Sri Prakasa, who had been given his old schoolboy job—incharge of the seating arrangements for the whole lecture series. Gandhi began provocatively by expressing his feelings of humiliation in having to address his audience in English rather than some vernacular, since the teaching of English in the Indian schools for the last fifty years had further kept the country from its objective of self-government. Drawing a contrast between the conditions of the poor in the filthy lanes around the temple and the gorgeous scene on the dais behind him, he turned to the massed princes and insulted them by impudently (he himself later said "humourously") calling attention to the jewels they were wearing as being "necessary for them to hold those treasures in trust for the nation before we could raise our ideals." Disregarding the stir caused by these remarks, he went on to comment ironically on the police precautions taken for the viceregal visit, and from there moved to the subject of anarchism and assassination which had long plagued the province and especially the students of Bengal.

Annie, upset already by his insult to the official guests of the university, many of whom had already sold some of their jewels to finance their gifts, listened with horror as she thought her ears told her how he was speaking of "bombs bringing about the annulment of the partition of Bengal, with praise for the heroism of those who threw them." When he declared flatly, "I am an anarchist," without explaining that he meant a "philosophical anarchist," she feared that a riot might be incited in that part of the "gunpowdery" audience which did not know his true beliefs. So when she heard an Englishmen behind her, whom she took for CID officer, whisper, "Everything he says is being taken down and will be sent to the Commissioner," she rose to a point of order and requested that it should be stopped which has become a purely political speech. The Maharaja, however, overruled her and gave him five minutes to complete his remarks. But by this time the sensitive princes were thoroughly roused. Gandhi charged later that when he started to explain the meaning of his previous statements he heard Mrs. Besant whispering to the princes just behind him that he was neither withdrawing his remarks nor explaining them. She, in turn, indignantly told the reporters that she was seated on the other side of the platform, nowhere near the princes, and said nothing to any of them. At any rate, she and the audience were startled to see the Maharaja of Alwar rise to his feet, ignoring the speaker, and move off the platform. As soon as the other princes perceived the manner of his protest, they too rose and flocked after him in a "Stampede", as a writer to the *Hindu* put it. Malaviya, adroitly trying to save the situation, gave the audience in a quick sentence the advice on the cowardliness of assassination that he was sure Gandhi had intended to give, announced the arrangements for the next day, and closed the meeting in spite of the cries of many of the remaining listeners of more proletarian status that Gandhi continue.

Then Annie saw Gandhi hurrying excitedly towards her as she was discussing the incident with a group still on the platform. She might have refrained from interrupting, he exclaimed. With a degree of warmth probably approximating his own, she answered: "How could we sit when you were compromising everyone of us on the platform? You ought not to have made the remarks you did." To the reporters who avidly followed up the affair for several days, Gandhi swore that he still did not know what she had taken such exception to.

Annie was bitter when she found that many more Indians were taking his side than hers. By 16 February the kettle was still boiling merrily. In an editorial, "Prophet and Politicians," she recalled a speech she had made back in 1894, in which she had discussed the three grades of "Nation-builders": the Prophet, the Populariser, and the Politician. Gandhi like Mazzini, she thought, was best fitted to be the first and highest of these, though the force of circumstances was making him a "Child of politics". Trying to cool the fire, she agreed with him that the "Benaras incident" was only one of those passing waves which always come and go. She and he, she protested, were both trying to serve "God and People" in their individual ways.

In another editorial she even suggested that Gandhi had fallen into "missionary hands" and was trying to turn the students against her. Gandhi of course indignantly denied having gone over to the missionarises, but the fact that a meeting of the Student's Hindu Association at the Pachaiyappa Hall in Madras was packed to suffocation while they listened to Mrs. Besant's talk on "Student's and Politics" would indicate that she had not lost all her influence on this group. Her point in this speech, to the effect that both sides of the episode had been thoroughly aired in the papers, that the public must now judge, and that she had great respect for Gandhi in spite of their differences, was re-emphasised in her final word on the unhappy affair, a signed editorial "In Defence of M.K. Gandhi." The editorial, which might equally well have been entitled "In Defence of Annie Besant," was a first-rate specimen of her favourite elevated line of Theosophical tolerance and forgiveness, which at the same time revealed her own pride and refusal, as if she could be ever wrong, though freely admitting that she could be. The hostility of some papers to her, she added in her favourite vein of martyrdom, was "so constant and so independent of anything she may do or leave undone, that it inflicts very little, if any, injury..."[2]

From an analysis and comparison of the conflicting accounts of this boisterous meeting, there would seem to have been two key passages in Gandhi's speech, one particularly involving Mrs. Besant, the other the princes. Perhaps if she had not been so upset by Gandhi's lateness and the consequent change in the order of the speeches, and had listened a little more carefully to what he had actually said about explaining the fallacy of the supposed relationship between bomb-throwing and partition, she would not have drawn the wrong conclusion about the direction of his remarks. On the other hand, Louis Fischer in the midst of his somewhat inaccurate account of the affair, quotes Gandhi as having declared, among some other quite incendiary remarks: "If we trust and fear God, we shall have to fear no one, not Maharajas, not Viceroys, not detectives, not even King George." Fischer also quotes Gandhi as having defended Mrs. Besant just before the Indian princes walked out, saying that she had acted thus because "she loves India so well, and she considers that I am erring in thinking audibly before you young men."

Although the Commissioner of Benaras sided with Mrs. Besant and threatened to take action against Gandhi for his behaviour, Pandit Malaviya dissuaded him in the name of the University. Even though the two main antagonists soon patched up a public peace and lauded one another for their services to India, and though later there were several occasions of actual co-operation between them, they always had mental reservations. Like Ramaswami Aiyar, V.K. Krishna Menon, who as another very promising young man had just then been discovered and aided by Mrs. Besant, told that he had always sensed an incompatibility of temperament between her and Gandhi. Even in the greatest people, Krishna Menon confessed that the personal element creeps in unawares, and human beings are always fallible and sometimes even small in their behaviour in the most serious situations. The "Benaras Affair" was one of the examples he cited.

Having thus left the Hindu University—temporarily—as a storm centre, Mrs. Besant returned to her more purely political labours. Forgetting her earlier Socialist doctrines, she got into a dispute at a lecture over taxes, export duties, and so on, and expressed her opinion that at the present time the advantages of state ownership of industries were dubious. She started a Gokhale Society in Madras, and travelled around India inaugurating branches and unveiling portraits of her idol. She was having trouble with the police at every turn. They were trying to dissuade people from buying her books, like *How India Wrought for Freedom* and *India, a Nation,* by going to their houses and asking if they had copies. And the government continued to create trouble for her by delaying or mutilating her mail and her telegrams. She had a premonition, unrelated to the Masters, that she was going to get into trouble with the Press Act, especially when papers like the Madras Mail seemed to be doing all they could in that direction by misquoting her in speeches on the relationship between government and education in India. There were disturbances and tensions in various schools in Madras, Calcutta, and smaller cities, which the English faculties did not seem to know how to deal with, except by repressions and even floggings, which in one instance had led to a student strike.

As she was driven closer and closer to her former *bete noire*, Tilak, she coined new political phrases. It was really a healthy fight between the "Yesterdays and Tomorrows" in the Home Rule movement, she cried. In May she went to Poona to deliver a big speech on "The Future of India". The whole city was a stir, and enthusiastic crowds exclusively "protected" by the police, turned out in spite of the great heat and the rains. As usual, she was the only European present. Tilak himself introduced her, saying that he had great faith in her sincerity, enthusiasm, and eloquence, and at the end of her series, in spite of her loss of voice, another well-known Poona leader, N.C. Kelkar, summed up topically by saying that Mrs. Besant and Tilak were the two "biggest radium atoms" now in India.

On 27th May the government itself exploded its bomb. The whole front page of *New India*, usually given over in the English fashion to classified advertisements carried the spearhead in the largest black type to be found: "Under The Press Act. Security Demanded From New India". In her leading article, "Bureaucracy in Excelsis," Mrs. Besant related what had happened. Just after her return from Poona and just before her departure for another engagement, but while she was in bed with a fever of 103° (Annie Besant could never resist these little personal touches), she had been served with a notice from the Acting Chief Presidency Magistrate, "dispensing with security in respect of the New India Printing Works," and requiring her to deposit Rs. 2000 with him as security within fourteen days. Yet no charges had been placed against her, and she was unable to learn of her sin. It would, in fact, she cried, be better for her if she were a criminal, since then we would get a trial. But "no law protects the citizen" under the Press Act of 1910.

Protests against the Government's action promptly began to pour in from all sorts of sources—from individuals, from organisations, and from the press. Even some of the most extremely reactionary, realizing the general threat to the sacred principle of freedom of the press, joined in the outcry. Encouraged, Annie decided that instead of refusing to pay security and being closed up at once, she would pay the money (it was probably Wadia's money; for the first time his name was now appearing in print as assistant editor) and keep on writing as she had been doing, so that on forfeiture she could go to the High Court and force the Government to show its hand. Thus she would have two more months after forfeiture, and attention would be drawn in England itself to the suppression of Free Speech in India. Four additional papers had now received similar notices.

All over India meetings of protest were organised, editorials were written, and money was raised. The staff of *New India* presented a letter to their "Dear Mother" from her "Humble Co-workers for India" assuring her of their devotion and making a contribution of Rs. 232. On 3 June she made a quick trip to interview the Governor of Calcutta, on an unspecified subject. Three days later she issued "An Appeal to H.E. the Viceroy," requesting him to investigate what was happening and to realise its widespread effect on Anglo-Indian relations.

On 12 June, in spite of the heat and the monsoons, the Victoria Hall and all its verandahs were filled to overflowing with both Indians and Whites at a protest meeting engineered by *New India*. There was no disorder, as even the Hindu admitted, and the police outside did not have to be called in. A resolution urging the repeal of the Press Act was passed by unanimous acclaim, and Annie was given an ovation. She had pointed out that the Government really had no need to fear a sixty-nine-year-old woman who had always abided by law, order, and constitutional means of reform. As she wrote confidently in the Commonweal, "...the crucifixion of today is the coronation of tomorrow." Prominent legislators like Wacha and Jinnah soon added their voices to the clamour. As the situation roared to a climax, Gandhi himself, after a dramatic caesura, joined in. The lowest citizen, who might never have heard of Annie Besant before, now had no excuse for not knowing her name.

While she waited impatiently for the forfeiture, she returned to her Home Rule campaign. She informed the *Madras Times* that she had never demanded Home Rule immediately, but only as a promised part of "Imperial Reconstruction" after the war. She also opposed the endorsement of the Moslem leader, Sir Ali Imam, for the Presidency of the Congress until he made his position on Home Rule clear. No one who would fail to push this movement much more strongly than Sinha had done, she insisted, should be chosen for this high office. She was no more than receiving her due, therefore, several of the provincial committees, like those in Bengal, Bihar and Madras, began placing her name on their lists of possible nominees. Early in July she reminded her readers that the Home Rule League, though not officially sanctioned by the last Congress, nevertheless still existed in skeleton form, and that those who belonged or wished to join for one rupee were asked only to send delegates pledged to Home Rule to the next Congress, to be held at Lucknow. Perhaps they would elect a President and form an organisation there. Groups were already coalescing. She and Tilak were working in amity; the Maharashtra League was working independently. Arundale, who long ago had given up any idea of returning to England as either tutor to Krishnamurti or as General Secretary to the English Section, had accepted the position of

Organising Secretary, and was already at work making speeches and signing up members. In London, C. Jinarajadasa was making addresses on the League at Lady Emily Lutyens's At Homes and similar places, though the *London Times* and the *Times of India* both ridiculed the H.R.L. in England as an "obscure organisation" of no consequence.

Though Mrs. Besant had long recognised the possibility of her internment, and wrote of it to Esther Bright, and the Englishman had recently come out in favour of interning more insurrectionaries than the two hundred who had already been so treated in Bengal, she was first to become the victim of the opposite process, that is, of externment from the Province of Bombay because "there are reasonable grounds for believing that Mrs. Annie Besant has acted and is about to act in a manner prejudicial to the public safety." The only reason for the action she could think of was the historical sketch introducing her How India Wrought for Freedom, which contained a savage indictment of England's past treatment of the country. The Advocate of India soon went so far as to urge that she be deported. Relatively few papers, however, defended the Governor's act. The Theosophists found the situation especially ironical since Lady De La Warr, a member of the Executive for the Home Rule for India League in England, was the sister of Lady Willingdon, the wife of the Governor, and both sisters were then in England together.

While Mrs. Besant was waiting to learn the next step of the Government, it announced that it was preparing to put Tilak on trial for orally spreading sedition and was binding him over for a year for a security of some Rs. 40,000 "to be of good behaviour". Some other Indian leaders had actually been arrested and jailed. The liberal English newspapers, like *Lansbury's Herald* and the *Manchester Guardian*, now began to take serious notice of these Indian developments.

As for Mrs. Besant's unrepressed ambition to be chosen President of the Congress, she could not make up her mind whether to withdraw her name or let it remain in the running. As new factors developed, she shifted from candidate to candidate, including herself, and finally declared that she would stay in and fight for India. The provincial committees, however, were not all equally impressed by her self-sacrifice, and when the necessary majority of five announced that they would vote for a veteran, but little-known, compromise candidate, Babu Amrika Charan Mazumdar, a philosopher and political historian, Mrs Besant asked her supporters at Lucknow as well as the committees of Bihar and the UP who had chosen her to shift their votes to him. The Madras Mail, which shortly before had vowed that it would make no further reference to her or her works, for from now on she must "pass, fade, and be forgotten" ("A remarkable career, of course; but finished"), found that it could not keep its word, and reported these and the following events.

At the end of August the Madras Parliament began its discussion of the first draft of a bill to establish Self-Government in India, under the name of "The Commonwealth of India Act". This bill, which was to be fought step by step and to get its foot at least into the doorway of the House of Commons,_was largely the work of Annie Besant and her close advisers. This "audacious"' proposal, cried *New India* in serialising the bill, represented the first time in Indian history that such thing had been attempted.

Knowing that the guillotine would ultimately fall, Annie kept her paper alive a little longer by forfeiting her security and then a new deposit (probably of Wadia's money) until she could appeal to the High Court. To be prepared for any eventuality, she now made a sort of political last will and testament, designating the various members of her entourage who would take over her several business and editorial functions if she was interned.

She could now devote her attention again to the Home Rule League which because of her persecution was expanding rapidly in spite of her promise to Congress to mark time for another year. S. Subramania Iyer was nominated as Honorary President, Annie Besant as active President, Ramaswami Aiyar and P. K. Telang as General Secretaries, Wadia as Treasurer, and Arundale as Organising Secretary. In the voting, Mrs Besant was backed by thirty branches, Tilak by one. Thus all the top offices, except that of Ramaswami Aiyar, were occupied by Theosophists, and even he was a strong sympathiser. The League, however, was far from a Theosophical organisation, and drew from all religious, racial, and political groups. Branches sprang up in Madras, Bengal, the Maharashtra, the UP, Bombay, and so on. By the end of October, it had sixty-six branches, though many of them were quite small. But no prominent Indians of the old school ever joined the H.R.L.[3] On 14 September she decreed the first annual celebration of Home Rule Day, and explained again why she preferred the term Home Rule to Self-Government, indicating that Self-Government outside the Empire was not aimed at.

There was, however, still enough rivalry between Annie Besant and B.G. Tilak for him not to be willing to fight under her banner, nor did his compact body of fighters want to battle under another. Consequently, in the same month of September he organised his own League; but with nice reciprocity he also joined her Home Rule for Indian League and she joined his Indian Home Rule League.[4] Many others, not feeling ready to discriminate in such a patriotic matter, likewise joined both. One of the double joiners was young Jawaharlal Nehru, who had already betrayed his strong inclination towards advanced positions by refusing to support Gokhale's Servants of India because it was, in his opinion, too unaggressive. He had also been shocked by a speech made by Sastri at a students' meeting at Allahabad, in which Sastri had advised discipline and submission to authority, and had practically told the boys to tattle on one another. At the outset, however, Nehru devoted his labours to Mrs Besant's group, until she too became too moderate for him, and he sought more extreme leadership.[5]

At the end of September her appeal under the Press Act came up before the High Court, and once more she argued her own case before the Advocate-General, the Public Prosecutor, the Government Solicitor, and the Chief Justice, maintaining that she had written nothing disloyal or illegal. While the Court's verdict was being awaited, she presented it with a second petition, to be argued by Ramaswami Aiyar, "alleging legal grounds against the imposition of the security on *New India*, and also raising the question of *ultra vires* with regard to the Press Act."[6] The proceedings were all very dull, full of technicalities and precedents and one observer saw at least six people asleep in court. The ultimate verdict was the usual one involving Mrs. Besant in any Indian court. Her revised application was dismissed, with their Lordships delivering separate but concurrent judgments, their consensus being that the order of the Local Government was illegal, that none of the offending articles was written in a spirit of disloyalty, but that the High Court was impotent to set the order aside. *The London Times* carried a brief Reuter's dispatch on the decision; both Reuter's and The Times had long wished to have as little to do with Mrs. Besant as possible.

Ruefully encouraged, Annie at once announced that she would take the case to the Appeal Court and if necessary to the Privy Council, London, to try to recover at least Rs. 12,000 of her Rs. 17.000. Of course, she failed in the Appeal Court also, and so she quickly cabled Graham Pole, who had helped her in a similar situation in the *Krishnamurti-Narayaniah case*, to come to Madras, where they could talk over their strategy before the Privy Council. He had already retained the Right Hon. Sir John Simon, K.C., M. P., as her leading counsel.

In the meantime her Bombay externment caused the Governor to be bombarded with petitions and memorials with hundreds of important and unimportant signatures requesting his reconsideration. Annie now began to change her mind about her popularity with the Indians and to write about "my personal influence and the love the people bear me" as the cause of the Government's hatred and fear. It made no difference to the bureaucrats: at the end of October she was externed also from the Central Provinces and Bihar.

But the new externment raised a new threat. She had planned for a long time to preside at a Theosophical Federation meeting at Amraoti. Now she could not go. She, therefore, penned an angry diatribe in *New India* on "Religious Liberty under the 'Defence' Act," pointing out the fresh menace to freedom when political prejudice was thus allowed to interfere with religion. Once more the press in general rallied to her side and the Press Association of India wired to express its gratitude to her for her "gallant fight". Her protest did no good, however, and she had to send Wadia to take her place and read her message. He not only turned out to be a tremendous success as a Theosophical speaker, but he made an equal hit when he defied the authorities by delivering an extra speech on the Home Rule League. The meeting had to be held out of doors, and large pictures of Mrs. Besant and Tilak hung over the platform.[7]

In the meantime, the Provincial Committees were meeting in preparation for the annual National Congress. The Bombay group were especially concerned with the situation because they felt that the honour of the state was involved. At their conference in October they elected as President the local Muslim lawyer and legislator, Mohamed Ali Jinnah, whose fame as a political leader at forty-two was already spreading over India. Known as the "Moslem Gokhale," he was at this time regarded as an emblem of Hindu-Moslem unity. Dressed in his checked trousers and black coat, with his black hair parted on one side and with his thick moustache, he concentrated everyone's attention on his fiery oratory. At this conference the new President spoke out boldly against the Press Act and Mrs Besant's exclusion from the province and found strong support from Gandhi, Sastri, Tilak and others.[8]

Fortunately Mrs. Besant had not been externed from Lucknow, where the Congress for 1916 was being held, and where she had seen to it that the T.S. convention was also to take place. As usual, she doubled back and forth between the two, and dropped in on some side meetings as well. She and the other "distinguished guests" were given a public reception at the railway station and then taken in procession through the main streets of the town. There was a record turnout of 2600 delegates under the new system of election. Though Mazumdar's address was wishy-washy and ineffective, in a signed article in *New India*, Mrs. Besant professed to be thoroughly content with both it and his programme, which she pretended was identical with her own, since he had come out clearly in favour of Home Rule. Her only speech was made in seconding Bannerji's Home Rule resolution, which this time was passed unanimously, with all present standing in their seats and shouting. The President was instructed to communicate the resolution to the Secretary of State for India, through the Viceroy.

The names of Annie Besant and Tilak were cheered.

The first Home Rule conference was held immediately afterward, with Mrs. Besant presiding over the thousand members who attended. She and Tilak spoke.

So far as the Indians were concerned, the Home Rule question was entering on the last stage of the three through which, according to A.P. Sinnett, a truth passes: (1) the Pooh-Pooh Stage—that of contempt; (2) the Bow-Wow Stage—that of fierce opposition; and (3) the Hands-All-Round Stage—"when, of course, everyone has always known it." So far as the British were concerned, however, the movement was still in its second stage, where it was long to stick.[9]

The acceptance of the doctrine of Home Rule by the Congress set off a series of national celebrations, though most of the English-edited newspapers in both India and Britain kept silence about them or sourly played them down as of no consequence. On the other side, an editorial in the *Paniabee* maintained that "Mrs. Besant's, services will live for all time in the recollection of a grateful country." When she immediately started off on a new propaganda tour for the League, she was feted at Calcutta, received with Vedic chants and camphor offerings at Vizianagram, lauded at Waltair and Cocananda, given a gaslight procession with fireworks and garlands at Rajamundry, and showered with flowers and fruits at Vellore. Everywhere there were huge crowds, procession and offerings. When, however, she returned to Madras at the end of January 1917 and prepared to participate in a new series of lectures at Gokhale Hall on political and governmental questions, she found that some of the "unco guid," as she called them, were refusing to use the hall at all—though it was the largest and most beautiful in Madras—simply because she had built it for the Young Men's Indian Association. In consequence of this discovery, she made an only partially facetious prophecy in the Commonweal, to the effect that probably after death a memorial meeting would be held for her there. This was one time when a prophecy by Annie Besant was more than realised, for a memorial meeting is held there for her on her birthday every year. On the other hand, she now began to prophesy with increasing certainty that she would not die before India became free.[10]

The Lucknow Congress saw a temporary rapprochement between the Moderates and the Extremists and the Congress-League Reforms Scheme was formulated. There was a strong statement for Indian Home Rule by Mrs. Besant in 1916 at the Cuddalore Political Conference. Lord Pentland who was then Governor of Madras sent for her and asked her to leave India. She declined. Thereupon in June 1917, Mrs. Besant as well as G.S. Arundale and B.P. Wadia, her colleagues in political and journalistic work, were interned under the orders of the Madras Government. Prior to her going into internment in Ootacamund she published an open letter in which she expressed her hope and firm belief that India would soon win Home Rule. A large number of protest meetings were held in Madras and all over India, and each succeeding month processions and meetings were organised. The internment was instrumental in strengthening the freedom movement not only in India but abroad. President Wilson of America, to whom a stirring letter was addressed by Sir Subramania Iyer, a retired High Court Judge and follower of Mrs. Besant, indicated to the British Government his sympathy with the Indian movement. A Proclamation was issued by the Secretary of State and the Viceroy, Lord Chelmsford, in 1917, which asserted the intention to grant self-government by the gradual development of self-governing institutions. In pursuance of this policy Mr. Montagu and the Viceroy toured the country to formulate a scheme after gathering public opinion. By then, Mrs. Besant was released as the result of continuous agitation and of growing public indignation, and in August 1917 she was made the President of the National Congress in Calcutta.

The opening words of her Presidential Address at Calcutta were these: "While I was humiliated, you crowned me with honour; while I was slandered, you believed in my integrity and good faith; while I was crushed under the heel of bureaucratic power, you acclaimed me as your leader, while I was silenced and unable to defend myself, you defended me and won for my release. It was proud to serve in lowest fashion, but you lifted me up and placed me before the world as your chosen representative. I have no words with which to thank you, no eloquence with which to repay my debt. My deeds must speak for me, for words are too poor. I turn your gift into service to the Motherland; I consecrate my life anew to her in worship by action. All that I have and am, I lay on the Altar of the Mother, and together we shall cry, more by service than by words, *Vande Mataram*!"[11]

Mrs. Besant went round the country pleading for Home Rule. The Montagu-Chelmsford proposals had instituted diarchy in the Provinces and left the Central Government untouched. One section of the Congress welcomed the reforms, whereas others denounced them. In August 1918 a special Congress was held at which Hassan Imam presided and the author was one of the Secretaries. At that time, the Moderates and the Extremists came to a kind of compromise and pleaded for full Responsible Government in the Provinces. In the Congress of December 1918, Mrs. Besant, agreeing that the Montagu-Chelmsford reforms were disappointing, proposed that they should be worked for what they were worth. In pursuance of this idea, she advised some of her friends to work from within the Government. It was then that some persons who had worked in the Congress and the Home Rule League accepted office under the Montagu-Chelmsford scheme; among them were Tej Bahadur Sapru, Surendranath Banerji, Chimanlal Setalvad, C. Y. Chintamani.

By 1918 Gandhiji had returned from South Africa having organised passive resistance, for redressing the grievances of British Indians. He laboured hard on behalf of the Indigo estate workers in Bihar and pleaded for his specific doctrines of *Ahimsa* or non-violence and *Satyagraha* or passive resistance. He developed doctrines which aimed at paralysing the Government by non-co operating with the functions of the Government, boycotting Legislative Councils and local bodies, calling upon lawyers to give up practice and exhorting people to court prosecution and disobey selected laws, suffering imprisonment and sale of their properties, and by conducting a campaign of non payment of taxes—all these methods being resorted to for compelling the Government to agree to *Swaraj*. Mrs. Besant objected on principle to Gandhiji's Civil Disobedience Movement though she carried on at the same time her intensive agitation for Home Rule. She, who had co operated with Bradlaugh in his movement of resistance to unjust laws but at the same time protested against violence, felt that she could not but take the attitude that she adopted with regard to Gandhiji's Civil Disobedience Movement. She at the same time expressed great admiration for Gandhiji's lofty ideals and personal self-abnegation and love of truth. Gandhiji's movement gained great popularity amongst the frustrated masses of the country. Mrs. Besant lost her popularity and, between the years 1920 and 1930, the Civil Disobedience Movement on the one hand and the Swaraj Movement on the other were both working in full force. In the course of the Civil Disobedience Movement many excesses were committed by the people; and, at one time, Gandhiji styled his own advice a "Himalayan blunder".

In 1919 there was a Committee of Enquiry appointed by Parliament and presided over by Lord Selborne. Mrs. Besant appeared as a representative of the National Home Rule League, and various other representatives pleaded the cause of India before the Committee. Before this date Mrs. Besant had founded the Madras Parliament in order to accustom political workers to parliamentary work and procedure. It met periodically in the Gokhale Hall in Madras which was erected at her own cost. The debates were published as memoranda, one of which was the Panchayat Act, and the other, a memorandum on Indian Finance. Steps were taken to frame a Commonwealth Bill. In the general elections devised under the Montagu-Chelmsford scheme, Mrs. Besant favoured entry into the legislature. Some members of the Congress Party themselves were dissatisfied with the results flowing from the policy of boycott of legislatures.[12]

The Swarajya Party was formed under the leadership of C.R. Das; and Pandit Motilal Nehru lent his support to it. Mrs. Besant, during the currency of the Montagu-Chelmsford scheme, went along with B.P. Wadia and D.K. Telang and Jamnadas, Dwarkadas and toured round England pleading for Home Rule. She started a branch of the Home Rule League in London with George Lansbury, M.P., as Chairman and John Scuff as Secretary and she was actively helped by persons like the Countess De la Warr, Lady Emily Lutyens, George Bernard Shaw, Mr. and Mrs Phillip Snowden and several others. Many Labour Associations and Trade Unions affiliated themselves to her League. She started a weekly newspaper in England called *United India*. She also published an overseas edition of *New India*.

A meeting attended by 60 members of Parliament was arranged in a Committee Room of the House of Commons at which addresses were given by V.S. Srinivasa Sastri of the Liberal Deputation, M.A. Jinnah of the Muslim League, Mrs. Besant as the representative of the National Home Rule League. Deputation and the author as that of the All India Home Rule League. A National Home Rule Deputation visited several cities and towns in England and Scotland and gave political lectures in support of Indian reforms. The following places were visited: London, Southport, Edinburgh, Glasgow, Stratford, Croydon, Battersea, Brighton, Preston, Liverpool, Wellesley, Wigan, Rochdale, Burnby, Bolton, Manchester, Oxford, Bradford, Harrogate, York, Newcastle, Stone, Cardiff, Bristol, Bath, Leeds and Harrow. Articles were written for several London newspapers, interviews were given to the Press, and generally every opportunity was taken to put Indians, case before the public. A deputation was received by Mr. Montagu, Secretary of State for India, at the India Office and there was a ladies' deputation organised by Sarojini Naidu, which included Mrs. Besant. A great meeting was held in Manchester at which Mrs. Besant spoke; a big meeting was held in the Queen's Hall, London, and a lecture in the Kingsway Hall, London, under the auspices of the Fabian Society in support of Indian reforms. Many questions were asked in Parliament by members belonging to the Parliament Committee whose work was to draw the attention of their local M.Ps. to any approaching Indian matter in the House of Commons and urge their attention to it and their votes on the side of reforms, or where needed, redress of wrong governmental action; to influence residents of the locality to write in a similar sense; to circulate information sent to them by the Central Committee; to correct untrue statements about India made in the local Press and write to it in favour of reforms and answer objections; and generally to give help and take any useful action in regard to urging on the British public their duty towards India. About 200 Trade Unions and Labour Organisations were affiliated to the Committee to help the agitation for India's freedom, to put pressure on members of Parliament to vote for India's freedom and Indian interests and to press parliamentary candidates to promise to support Indian freedom and Indian interests. John Scurr took a most prominent part in this endeavour. Meetings were organised in England by several women's societies, the "Britain and India League" and various Labour Organisations. Mrs. Besant addressed the famous Pioneer Club, believed to be the first women's society, a Labour Conference at Southport and a meeting of the Home Rule for Scotland League.

In 1919 the Amritsar Congress was held and it marked a great step forward in the Indian agitation. The police excesses in the Jallianwala Bagh in Amritsar and other events during that year created a terrible unrest in India.

Between the years 1922 and 1924, Mrs. Besant, in the course of the work of the Madras Parliament, framed in consultation with her associates a Swaraj Constitution for India in accordance with the pattern adopted by the several colonies in the British Empire which put forward a draft Constitution in the course of their campaign for attaining political freedom. This Constitution was put forward before a National Conference in Delhi convened amongst others by Tej Bahadur Sapru, Sivaswami Aiyar, Srinivasa Sastri, Purushotamdas Thakurdas and H.S. Gour with Mrs. Besant as Secretary. This Conference outlined the essentials of a free Constitution. Mrs. Besant, during the years 1924 and 1929, organised a series of Conferences with the help of men like Sapru and V.S. Srinivasa Sastri, published several pamphlets on self-determination and self-government and the objects of the National Conference and sent out various bulletins over her signature, acting on behalf of these Conferences. In the meantime, the Conservative Government had been defeated in the general elections in England; and the Labour party that came into power through its majority was very insecure.

A noteworthy effort was by a National Convention which came into existence in 1924 at Allahabad. A deputation to England was appointed consisting of Mrs. Besant, Lady Emily Lutyens, Sastri and Munshi Iswar Saran. This deputation presented a memorandum to the India Office, making a demand for Dominion Home Rule, signed by Mrs. Besant, Sastri, Munshi Iswar Saran, Ali Imam, K.G. Gupta and several others; and it suggested that India should draft her own Constitution as was done by the Dominions and the Irish Free State. A series of meetings were held and a public campaign in England was opened by Bradford. On the 23rd of July 1924, there was a great demonstration in the Queen's Hall in honour of Mrs. Besant's 50th anniversary of entry into public life. Men like Ramsay MacDonald, Lord Haldane, Philip Snowden, Lord Baden-Powell and others associated themselves with the demonstration and with her efforts, on behalf of India.

With this background, Mrs. Besant came back to India and made an appeal for a united front of the Congress, the Home Rule League, the Swarajya Party, Liberal Federation, the Muslim League and the Independent Party. When she approached Gandhiji, he replied that the Congress Constitution had broken down as also the non-cooperation and civil-disobedience movements, and therefore, he suggested a programme which was considered appropriate and efficacious, including concentration on the spinning wheel, on removal of untouchability, and unity among all communities. Mrs. Besant, though she did not agree with Gandhiji's programme, agreed to spin for half an hour a day if non-cooperation and civil disobedience were suspended. But at the Belgaum Congress, non-cooperation and the total rejection of the Montagu-Chelmsford Reforms were decided upon. Thereafter, Mrs. Besant arranged for a United Conference and, after another National Convention, the Commonwealth of India Bill which was framed mainly on the initiative of Mrs. Besant was put into parliamentary form with the assistance of the British Attorney-General, Mr. Slesser.

The Bill was presented by Mr. George Lansbury, M.P., in the British House of Commons and was supported by Snell, Col. Wedgwood, Charleston, Haden Guest, Waltex Baker, Gillet, Lees Smith, James Hudson, Dalton and Scurr, and was ordered to be printed by the House of Commons on December 17, 1925. It passed through the First Reading in Parliament. But some of the Congress leaders cabled to England that it was not their Bill. It lapsed in the subsequent session of the British Parliament. In 1929, Mrs. Besant again went to England and arranged for another series of meetings and appeals to England to confer Home Rule on India.[13]

It was on July 2, 1928, that Ramsay MacDonald made the following statement: "I hope that within a period of months, rather than of years, there will be a new Dominion added to the Commonwealth of Nations; I refer to India."

On June 27, 1929, there was a grand demonstration in Free Trade Hall, Manchester on "Britain's Rule in India—A danger to the peace of the World", organised by the Commonwealth of India League, Manchester, and presided over by George Lansbury, a Cabinet Minister in which Mrs. Besant, Munshi Iswar Saran, B. Siva Rao, Major Graham Pole and others spoke. It produced a great mental upheaval in England. There was also a "Commonwealth of India" Conference held at Caxton Hall, London, at which Mrs. Besant presided. Many Indians, including Munshi Iswar Saran and others took part. A resolution was passed unanimously by the delegates present in favour of "Dominion Status" for India. On June 29, 1929, Mrs. Besant gave a public lecture on the subject of "Dominion Status for India".

There was an All Parties Conference held in Bombay joined by different political parties which appointed a committee to discover ways to draw parties together, to bring about Hindu-Muslim unity and outline a scheme of Swaraj.

Between 1925 and 1929, Mrs. Besant was very active in endeavouring to bring different political parties and leaders together for framing an agreed scheme of Swaraj and creating public opinion to that end, by her articles in the Press, by lectures in different parts of the country and by circulation of literature. A scheme of Swaraj was ultimately drawn up under the guidance of Pandit Motilal Nehru. An All Parties Conference was held at Calcutta in December 1929. Jinnah, representing the Muslim League, was there. M.R. Jayakar was also there, championing the cause of the Hindus. After a few clauses in the scheme were passed, it foundered when it came to the stage of fixing the percentage of Muslim representation in the Central Legislature. Mr. Jayakar, speaking for the Hindus, limited the percentage to 25 and Mr. Jinnah claimed 30 per cent for Muslims. Mrs. Besant was not particular whether it was 25 or 30 per cent because, after all, 30 per cent was only a minority and would not make any basic difference. The Muslim League's claims were subsequently improved upon and the one point of difference developed into 14 points of difference which subsequently became bitter points of controversy and conflict in the subsequent Round Table Conferences, which were held in London and at which Gandhiji alone represented the Indian National Congress. The Simon Commission was later constituted by the British Government with a wholly British personnel under the Chairmanship of Lord Simon. This body was boycotted both by the Indian National Congress and the National Home Rule League. In spite of differences between the political parties, which stood in the way of framing a Swaraj Constitution for India by Indians, Mrs. Besant continued to carry on her campaign with unabated vigour through the Press, the platform and political organisations.

REFERENCES

1. The following account is pieced together from *New India*, 3-19 February 1916; *Commonweal*, 4-11 February 1916; J. Nehru, *Autobiography* (London: The Bodley Head), 1967, p. 533.; *Messenger*, April 1916; Prakasa, *Annie Besant, op. cit.*, pp. 148-51; Morton, pp. 118-19; Louis Fischer, Gandhi (Stuttgart: Tauchinty), 1953, pp. 133-37.
2. Nethercot, Arthur, H., *The Last Four Lives of Annie Besant, op. cit.*, p. 246.
3. Fischer, Louis, *Gandhi, op. cit.*, pp. 136-38.
4. Diwakar, Das, "M. A. Jinnah", *Pakistan Standard*, 24 December 1954.
5. Sri Prakasa, *Annie Besant, op. cit.*, p. 241.
6. Nehru, J., *Autobiography, op. cit.*, pp. 30-3
7. *New India*, 7 June 1917.
8. Diwakar, Das, "M. A. Jinnah", *Pakistan Standard*, 24 December 1954.
9. *New India*, 19 October 1916.
10. *Commonweal*, 24 March 1916.
11. *Ibid.*, 13 July 1917.
12. Aiyar, C.P. Ramaswami, *Annie Besant, op. cit.*, p. 65..
13. *Ibid.*, p. 69.

BIBLIOGRAPHY

Athalye, D.V., *The Life of Lokmanya Tilak* (Poona: Sadashiv Path), 1921.

Aiyar, C.P. Ramaswami, *Annie Besant* (New Delhi: Publications Division, Ministry of Information and Broadcasting, Government of India), 1963.

Banerji, Arun, *Finances in the Early Raj: Investments and the External Sector* (New Delhi: Sage Publications), 1995.

Besant, Annie, *A Bird's Eye View of India's Past as the Foundation for India's Future* (Adyar: Theosophical Publishing House), 1931.

Besant, Annie, *Autobiography* (London: J.M. Dant), 1893.

Besant, Annie, *Civil and Religious Liberty* (Adyar: Theosophical Publishing House), 1883.

Besant, Annie, *How India Wrought for Freedom* (Adyar: Theosophical Publishing House), 1915.

Besant, Annie, *Lectures on Political Science* (Adyar: Theosophical Publishing House), 1918.

Besant, Annie, *Popular Lectures on Theosophy* (Adyar: Theosophical Publishing House), 1883.

Besant, Annie, *Shall India Live or Die?* (Adyar: Theosophical Publishing House), 1925.

Besant, Annie, *The Ancient Wisdom: An Outcome of the Theosophical Feelings* (Adyar: The Theosophical Publishing House), 1955.

Besant, Annie, *The Changing World* (Adyar: Theosophical Publishing House), 1909.

Besant, Annie, *The Future of Indian Politics* (Adyar: Madras), 1922.

Besant, Annie, *The Great Plan* (Adyar: Theosophical Publishing House), 1920.

Besant, Annie, *The Ideals of Future of Theosophy* (Adyar: Madras), 1912.

Besant, Annie, *The Riddle of Life* (Adyar: The Theosophical Publishing House), 1955.

Black, R.D., *Collision, Economic Thought and the Irish Question, 1817-1870* (Cambridge: Cambridge University Press),1960.

Blaug, Mark, *Economic Theory in Retrospect* (Cambridge: Cambridge University Press), 1978.

Bury, J.B., *History of the Freedom of Thought* (London: Oxford University Press), 1960.

Bury, J.B., *The Idea of Progress* (London: Oxford University Press), 1960.

Chakarverty, S., "Drain Theory of Dadabhai Naoroji," a paper presented to the Indian Economic Association Conference, December, 1968.

Chandra, Bipin, *Ranade's Economic Writings* (New Delhi: Gian Publishing House), 1990

Chandra, Bipan, *The Rise and Growth of Economic Nationalism in India, Economic Policies of Indian National Leadership 1880-1905* (New Delhi: People's Publishing House), 1966.

Charlesworth, Neil, *British Rule and the Indian Economy, 1800-1914* (London: Macmillan), 1982.

Chaudhuri, K.N., "India's International Economy in the Nineteenth Century: An Historical Survey", *Modern Asian Studies*, Vol. 2, No.1 (January), 1968.

Chintamani, C.Y., *Indian Social Reforms*, Part II (Allahabad: Kitabistan), 1947.

Clark, Colin, *Conditions of Economic Progress* (London: MacMillan and Co. Ltd.), 1949.

Dantwala, M.L., *Poverty in India, Then and Now, 1870-1970* (Delhi: MacMillan), 1973.

Das, M.N., *Indian National Congress Versus the British*, Vol. 1, (1885-1918), Delhi, 1978.

Dasgupta, Ajit K., *A History of Indian Economic Thought* (London: Routledge), 1993.

Datta, Bhabatosh, *The Evolution of Economic Thinking in India* (Dr. P.N. Banerjea Memorial Lectures, 1961) (Calcutta: Federation Hall Society), 1962.

Dimand, Robert W., "I have no Great Faith in Political Arithmetick: Adam Smith and Quantitative Political Economy" in *Measurement, Quantification and Economic Analysis: Numeracy in Economics* edited by Ingrid H. Rima (London: Routledge), 1995.

Dutt, Romesh, *The Economic History of India Under the British Rule*, Vols. 1,2 & 3 (New Delhi: Publications Division, Ministry of Information and Broadcasting, Government of India), 1963.

Gadgil, D.R., *Industrial Evolution of India* (Delhi: Oxford University Press), 1959.

Ganguli, B.N., *Indian Economic Thought: Nineteenth Century Perspectives* (New Delhi: Tata McGraw-Hill Publishing Co. Ltd.), 1978.

Georgescu-Roegen, Nicholas, *Analytical Economics: Issues and Problems* (Cambridge, Massachusetts: Harvard University Press), 1966.

Gidumal, Dayaram, *The Status of Women in India* (Ahmedabad: Navajivan Publishing House), 1953.

Gopalakrishnan, P.K., *Development of Economic Ideas in India, 1880-1914* (Gravenhange: Institute of Social Studies), 1954.

Green, T.H., *Principles of Political Obligations* (London: Allen and Unwin), 1934.

Gupta, Ajit Das, *A History of Indian Economic Thought* (Madras: T.R. Publications Private Limited), 1990.

Hallowell, G.H., *Modern Political Thought* (New York: Holt), 1953.

Jagirdar, P.J., Mahadeo Govind Ranade (New Delhi: Publications Division, Ministry of Information and Broadcasting, Government of India), 1971.

Jagirdar, P.J., *Studies in the Social Thought of M.G. Ranade* (New York: Asia Publishing House), 1963.

Javedkar, S.D., *Adhunik Bharat* (Pune: Sadashiv), 1947.

Jha, Shiva Chandra, *A History of Indian Economic Thought* (Calcutta: Firma K.L. Mukhopadhyay), 1981.

Kabir, Humayun, *The Indian Heritage* (Bombay: Bharatiya Vidya Bhawan), 1953.

Karmakar, A.D., *Economic Thought of Modern India* (Dharwar: J.S.S., Institute of Economic Research), 1974.

Karnonakaran, K.P., *Indian Political Thought 1832-1921* (New Delhi: Gitanjali Publishing House), 1996.

Karunakaran, K.P., *Indian Politics from Dadabhai Naoroji to Gandhi: A Study of the Political Ideas of Modern India* (New Delhi: Gitanjali Publishing House), 1988.

Kellock, James, *Mahadev Govind Ranade: Patriot and Social Servant* (New Delhi: Rupa & Co.), 2003.

Knight, Robert, *India: A Review of England's Financial Relations Therewith* (London: W.J. Johnson), 1968.

Lakatos, Imre, "Falsification and the Methodology of Scientific Research Programmes" in *Criticism and the Growth of Knowledge* edited by Imre Lakatos and Alan Musgrave (Cambridge: Cambridge University Press), 1970.

Leadbeater, C.W., *Man Visible and Invisible* (Adyar: The Theosophical Publishing House), n.d.

Lipton, Michael and Ravallion, Martin, "Poverty and Policy", *Policy Research Working Papers*, No. WPS 1130 (Washington: The World Bank), 1993.

Loganathan, *A History of Economic Thought* (New Delhi: S. Chand and Company Ltd.), 1987.

Madan, G.R., *Economic Thinking in India* (Madras: S. Chand & Co.), 1982.

Majumdar, Jatindra, Kumar (ed.), *Indian Speeches and Documents on British Rule 1821-1918* (London), 1937.

Mankar, G.A., *A Sketch of the Life and Works of the Late Mr. Justice M.G. Ranade* (Bombay: Dayabhai), 1902, Vol. I.

Masani, R.P., *Builders of Modern India - Dadabhai Naoroji* (New Delhi: Publications Division, Ministry of Information and Broadcasting, Government of India), 1960.

Masani, R.P., *Dadabhai Naoroji: The Grand Old Man of India* (London: George Allen and Unwin Ltd), 1939.

Marx, Karl and Fredrick Engels, *On Colonialism* (Moscow: Progress Publishers), 1965.

Mukherjee, M., *National Income of India Trends and Structure* (Calcutta: Statistical Publishing Society), 1969.

Nabers, Lawrence, 'The Positive and Genetic Approaches' in *The Structure of Economic Science Essays on Methodology,* edited by Sherman Roy Krupp (Englewood Cliffs, N.J.: Prentice-Hall, Inc.), 1966.

Nanda, B.R., *Gokhale, Gandhi and the Nehru's Studies in Indian Nationalism* (London: George Allen and Unwin Ltd.), 1990.

Naoroji, Dadabhai, *Poverty and Un-British Rule in India* (New Delhi: Publications Division, Government of India), 1962.

Naoroji, Dadabhai, *India's Wrongs and Remedy* (Madras: The Commonwealth Office), 1917.

Nethercot, Arthur, H., *The Early Four Lives of Annie Besant* (London: Rupert Hart Davis), 1961.

Nethercot, Arthur H., *The Last Four Lives of Annie Besant* (London: Rupert Hart Davis), 1963.

Olcott, H.S., *Old Diary Leaves* (Adyar: Theosophical Publishing House), 1975.

Owen, H.P., "The Home Rule Leagues, 1915-18", in Low, D. A. (ed.), *Soundings in Modern South Asia History* (London: Oxford University Press), 1968.

Pandey, Shridhar, *Economic History of India* (Hindi) (Patna: Motilal Banarsidars), 1991.

Panikkar, K.M., *Asia and Western Dominance* (London: George Allen and Unwin), 1964.

Pound, Roscoe, *Spirit of Common Law* (New York: Yale University Press), 1955.

Parekh, C.L, *Essays, Speeches, Address and Writings (on Indian Politics) of the Hon'ble Dadabhai Naoroji* (Bombay: Asia Publishers), 1887.

Parliamentary Debates (House of Commons) Fourth Series Vol. XI, 13th April 1893.

Paravate, T.V., *Mahadev Gobind Ranade* (Bombay: Asia Publishing House), 1963.

Patwardhan, R.P. (ed.), *Dadabhai Naoroji Correspondence, Child and Youth Section*, Vol. II, Part I (Bombay: Allied Publishers), 1977.

Patwardhan, W.B., "Justice M.G Ranade and Social Reform in Western India," *Indian Interpreter*, January 1911.

Paul, R.R., *History of Economic Thought* (New Delhi: Kalyani Publishers), 1979.

Ranade, Ramabai, *Amchya Ayushatil Kahil Athavani* (trnsl.) by Kusumavati Deshpande, Ranade His Wife's Reminiscences (New Delhi: Publications Division, Ministry of Information and Broadcasting), 1963.

Ranade, Mahadev Govind, *Ranade's Economic Writings* edited by Bipan Chandra (New Delhi: Gian Publishing House), 1990.

Ranade, M.G., *Religious and Social Reforms* (New Delhi: Sahitya Akademi), 1992.

Ranade, M.G., *Rise of the Maratha Power* (New Delhi: Publications Division, Ministry of Information and Broadcasting, Government of India), 1961.

Ranade, Ramabai, *Miscellaneous Writings of the Late Hon'ble Mr. Justice M.G. Ranade* (New Delhi: Sahitya Akademi), 1992.

Rao, V.K.R.V., *An Essay on India's National Income, 1925-1929* (London: George Allen and Unwin), 1939.

Robinson, Joan, *An Essay on Marxian Economics* (London: Macmillan), 1966.

Schumpeter, J.A., *History of Economic Analysis* (New York: Oxford University Press), 1954.

Scweinitz Jr., Karl de, *The Rise and Fall of British India: Imperialism as Inequality* (London: Methuen & Co.), 1983.

Sen, J.M., *History of Elementary Education in India* (Calcutta: Superintendent, Government Printing), 1917.

Sen, R.K. and Mandal, S.K., *Ranade's Analysis of India's Population Problem and the Views of Naoroji and Doskar*, I.E.A., 75th Conference Volume, 1992.

Sen, Sunânda, *Colonies and the Empire India, 1890-1914* (Hyderabad: Orient Longman), 1922.

Shanmugasundaram, V., *Indian Economic Thought and Policy* (New Delhi: S. Chand and Company Ltd.), 1981.

Singh, Gurumukh N., *Indian Constitutional and National Development* (Ajmer: Sachin), 1959.

Singh, V.B., *From Naoroji to Nehru* (New Delhi: MacMillan), 1975.

Sri Prakasa, *Annie Besant* (Bombay: Bharatiya Vidya Bhavan), 1954.

Speeches and Writings of Gopal Krishan Gokhale (Bombay: Asia Publishing House), 1962.

Stark, Werner, *The History of Economics in its Relation to Social Development* (London: Kegan Paul, Trench, Trubner & Co.), 1944.

Studenski, Paul, *The Income of Nations* (New York: New York University Press), 1958.

Tahmankar, D.V., *Lokmanya Tilak* (London: George and Allen), 1956.

Tarachand, *History of Freedom Movement in India*, Vol. III (New Delhi: Publications Division, Ministry of Information and Broadcasting, Government of India), 1972.

Viner, Jacob, *Studies in the Theory of International Trade* (London: George Allen and Unwin), 1960.

Watkins, F., *The Political Traditions of the West* (London: Nelson), 1953.

Williams, *Gertrude Mervin, The Passionate Pilgrim* (New York), 1931.

World Bank, *World Development Report, 1990* (Oxford: Oxford University Press), 1990.

Zaidi, Moin A., *Dadabhai Naoroji, Speeches and Writings* (New Delhi: Indian Institute of Applied Political Research), Vol. 1, 1985.

Index